OF BOOKS AND BOTANY IN EARLY MODERN ENGLAND

Literary and Scientific Cultures of Early Modernity

Series editors:

Mary Thomas Crane, Department of English, Boston College, USA
Henry Turner, Department of English, Rutgers University, USA

This series provides a forum for groundbreaking work on the relations between literary and scientific discourses in Europe, during a period when both fields were in a crucial moment of historical formation. We welcome proposals for books that address the many overlaps between modes of imaginative writing typical of the sixteenth and seventeenth centuries – poetics, rhetoric, prose narrative, dramatic production, utopia – and the vocabularies, conceptual models, and intellectual methods of newly emergent "scientific" fields such as medicine, astronomy, astrology, alchemy, psychology, mapping, mathematics, or natural history. In order to reflect the nature of intellectual inquiry during the period, the series is interdisciplinary in orientation and publishes monographs, edited collections, and selected critical editions of primary texts relevant to an understanding of the mutual implication of literary and scientific epistemologies.

Other titles in the series:

Milton's Secrecy
And Philosophical Hermeneutics
James Dougal Fleming

Poetry and Ecology in the Age of Milton and Marvell
Diane Kelsey McColley

Science, Literature and Rhetoric in Early Modern England
Edited by Juliet Cummins and David Burchell

Biblical Scholarship, Science and Politics in Early Modern England
Thomas Browne and the Thorny Place of Knowledge
Kevin Killeen

Robert Greene's Planetomachia *(1585)*
Edited by Nandini Das

Of Books and Botany in Early Modern England

Sixteenth-Century Plants and Print Culture

LEAH KNIGHT
Brock University, Canada

Routledge
Taylor & Francis Group
LONDON AND NEW YORK

First published 2009 by Ashgate Publishing

Published 2016 by Routledge
2 Park Square, Milton Park, Abingdon, Oxon OX14 4RN
711 Third Avenue, New York, NY 10017, USA

Routledge is an imprint of the Taylor & Francis Group, an informa business

British Library Cataloguing in Publication Data
Knight, Leah
Of books and botany in early modern England: sixteenth-century plants and print culture. – (Literary and scientific cultures of early modernity)
1. English literature – 16th century – History and criticism 2. Horticultural literature – History and criticism 3. Botany in literature 4. Gardens in literature
I. Title
820.9'364'09031

Library of Congress Cataloging-in-Publication Data
Knight, Leah
Of books and botany in early modern England: sixteenth-century plants and print culture / by Leah Knight.
p. cm. — (Literary and scientific cultures of early modernity)
Includes bibliographical references.
ISBN 978-0-7546-6586-1 (hardcover: alk. paper)
1. English prose literature—Early modern, 1500–1700—History and criticism. 2. Literature and science—England—History—16th century. 3. Botany—History—16th century. 4. Early printed books—England—History—16th century. 5. England—Intellectual life—16th century. 6. Turner, William. d. 1568. 7. Gerard, John, 1545–1612. 8. Botany in literature. I. Title.

PR767.K57 2009
828'.30809364—dc22

2008029298

ISBN 9780754665861 (hbk)

For my parents

Contents

List of Illustrations *viii*
Writing on Hyacinths: A Prelude *xi*
Acknowledgments *xvii*

1 Published Virtues of the Earth: An Introduction 1

2 The Bookish Nature of Botanical Culture: Continental Contexts 13
Botanical Renaissance: A Potted History 15
Words and Things in Humanist Herbalism 17
Botany by the Book 20
Botanical Gardens in Book Form: The Herbaria 28

3 Botanical Reformation in William Turner's Books of Nature 39
Biographical Contexts 41
Anthological Botany 46
Turner's Reforms: Naming and Translating 49
What's in a Name? Truth, Falsity, and Fiction 58

4 John Gerard's Uncommon Ground 69
The Garden Grows: From Catalogue to Herbal 71
The Anthology of Nature: Gerard's Herbal Authorship 75
Botanical Poetics and the Rhetoric of Description 83
Imaginary Gardens: Poets as Herbalists 99

5 Domesticated Plants and Domesticating Books: Cultivating Household Textual Collection 111
The Place of Books 113
The Place of Plants 115
Gallery and Paradise 119
Stillroom and Study 124
Posies Transposed 129

Epilogue 133

Bibliography *135*
Index *153*

List of Illustrations

0.1 A two-page spread from John Gerard's *The Herball or Generall Historie of Plantes* (1597), reproduced by permission of The Huntington Library, San Marino, California xii–xiii

0.2 An illustration from Geoffroy Tory's *Champ Fleury* (1529), reproduced by permission of The Huntington Library, San Marino, California xv

1.1 The title page of Thomas Howell's *The Arbor of Amitie* (1568), by permission of the Bodleian Library, University of Oxford, 8°H. 4.4 (2) Art Seld 2

1.2 The title page of John Maplet's *A Greene Forest* (1567), reproduced by permission of The Huntington Library, San Marino, California 3

2.1 One of David Kandel's illustrations in Hieronymus Bock's *Kreuterbuch* (1551), reproduced by permission of The Huntington Library, San Marino, California 26

2.2 Portraits of the illustrators of Leonhart Fuchs's *De Historia Stirpium* (1542), by permission of the Osler Library of the History of Medicine, McGill University, Montreal, Quebec, Canada 27

2.3 A page from the herbarium of the apothecary Petrus Cadé. Reproduced by permission of the Utrecht Herbarium and the Petrus Cadé Herbarium website (http://www.bio.uu.nl/~herba/Cade/). Photograph by Renske Ek 30

2.4 The J. Paul Getty Museum, Los Angeles, Joris Hoefnagel and Georg Bocskay, Maltese Cross, Mussel, and Ladybird, Fols. 1–129 written 1561–62, illumination added about 1591–96. Watercolour, gold and silver paint, and ink on parchment and paper bound between pasteboard covered with red morocco, 16.7 x 12.4 cm 34

2.5 The J. Paul Getty Museum, Los Angeles, Joris Hoefnagel and Georg Bocskay, Trompe l'Oeil Stem of a Maltese Cross, Fols. 1–129 written 1561–62, illumination added about 1591–96, Watercolour, gold and silver paint, and ink on parchment and paper bound between pasteboard covered with red morocco, 16.7 x 12.4 cm 35

2.6 The J. Paul Getty Museum, Los Angeles, Attributed to Bernard Palissy, Oval Basin, about 1550, Lead-glazed earthenware, 47.9 x 36.8 cm 36

4.1 A two-page spread from John Gerard's *The Herball or Generall Historie of Plantes* (1597), reproduced by permission of The Huntington Library, San Marino, California 101

5.1 An illustration from Thomas Hill's *The Gardener's Labyrinth* (1594), reproduced by permission of The Huntington Library, San Marino, California 121

Writing on Hyacinths: A Prelude

John Gerard was a sixteenth-century English surgeon, gardener, and herbalist. He is remembered primarily for his central role in the publication of a printed herbal, or encyclopedia of plants, in 1597. I would like to consider a single excerpt from his book as a prelude to my own. This excerpt exemplifies several of the unexpected ways in which a culture of plants and a culture of texts met in the early modern world. In a chapter on the English hyacinth, Gerard alludes to a now obscure material connection between books and plants. He also demonstrates some of the difficulties of botanical naming before the establishment of a standardized nomenclature. With a straight face, he discusses the idea of a plant that incorporates writing within its body. Finally, he integrates poetic accounts seamlessly into his botanical description. A close reading of this example thus offers a brief overview of many of the meaningful interstices between plants and writing that were made with remarkable intricacy in his day.

Printed books, then as now, were made from plants at the most basic material level of the botanical pulp—then flax, now softwood—that formed their pages. But as an incidental part of his description of the English Hyacinth, Gerard reveals that this was not the only material dependence of books upon plants. The plant's root, we learn, "is Bulbus, ful of a slimy glewish iuice, which will serue to set feathres vpon arrows in steed of glew, or to paste bookes with ..." (99). Plants and pages were thus quite literally bound together. While the botanical composition of a book's pages might be more apparent to a maker of paper than to the ordinary buyer, Gerard's tip suggests that even ordinary readers and writers may have been familiar with a material interface between paper and plants in this now quite unfamiliar way.

Gerard appears to draw his information about the adhesive quality and use of hyacinth juice directly, though without acknowledgment, from the work of his most significant English predecessor in herbalism, William Turner. In the second part of his own herbal, first printed in 1562, Turner had written under "The virtues of Hiacinthus" that "[t]he boys in Northumberland scrape the root of the herb and glue their arrows and books with that slime that they scrape off' (398). Turner, who grew up in Northumberland, is almost certainly drawing on his own childhood experience as a source for this botanical fact; Gerard's apparent borrowing of Turner's remark is thus an example of the sometimes covert anthological work that was so prevalent in herbals of the period. Botanical writers like Gerard thereby participated in a larger literary culture of gathering textual fragments and reproducing them, with or without acknowledgment, in other texts. Turner's remark becomes, through its transference to Gerard's herbal, a kind of commonplace: a piece of text transplanted from its original site and reproduced in another. Herbals resemble

100 THE FIRST BOOKE OF THE

6 The white Engliſh Iacint is altogither like vnto the precedent,ſauing that the leaues hereof are ſomwhat broader, the flowers more open,and very white of colour.

There is found wilde in many places of England, another ſort, which hath flowers of a faire carnation colour, which maketh a difference from the other.

✲ *The place.*

The blew Harebels do growe wilde in woods, copſes, and in the borders of fields euery where through England.

The other two are not ſo common,yet do they grow in the woods by Colcheſter in Eſſex, in the fieldes and woods by Southfleete neere vnto Graues-end in Kent, as alſo in a peece of grounde by Canturburie called the Clapper, in the fields by Bathe, about the woods by Warrington in Lancaſhire and other places.

✲ *The time.*

They flower from the beginning of Maie vnto the end of Iune.

✲ *The names.*

The firſt of our Engliſh Iacints is called *Hiacinthus Anglicus,* for that it is thought to grow more plentifully in England than elſe where: of *Dodoneus Hyacinthus non ſcriptus,* or the vnwritten Iacint.

The other *Hyacinthus Belgicus candidus,* or the Lowe countrey Iacint with white flowers.

7 *Hyacinthus orientalis cæruleus.*
The blew Orientall Iacint.

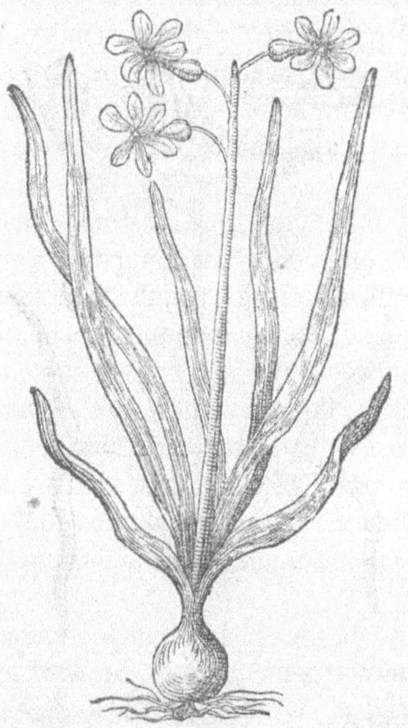

8 *Hyacinthus orientalis Polyanthos.*
Double Orientall Iacint.

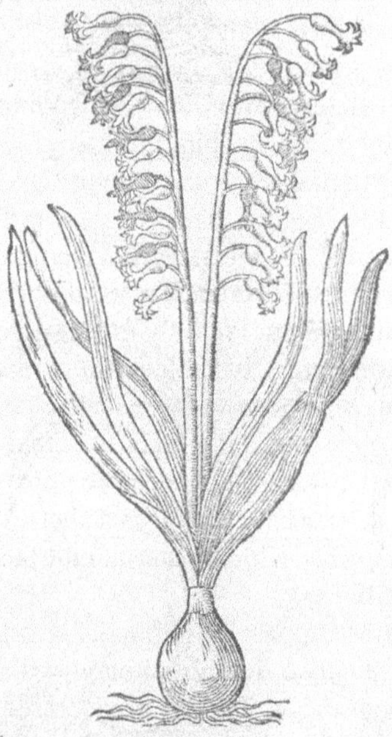

✲ *The deſcription.*

7 The orientall Iacint hath great leaues,thicke,fat,and full of iuice, deepely hollowed in the middle like a trough: from the middle of thoſe leaues riſeth vp a ſtalke two hands high, bare without leaues,very ſmooth, ſoft and full of iuice, loden toward the top with many faire blew flowers, hollow like a bell, greater than the Engliſh Iacint. The roote is great, Bulbus or Onion faſhion, couered with many ſcalie reddiſh filmes or pillings,ſuch as thoſe that couer Onions.

8 The double Iacint or Iacint with many flowers (for ſo doth the word *Polyanthos* import) hath very many large and broad leaues, ſhort and very thicke, fat or full of ſlimie iuice: from the middle whereof riſe vp ſtrong thicke groſſe ſtalks, bare and naked, ſet from the middle to the top with many

Fig. 0.1 A two-page spread from John Gerard's *The Herball or Generall Historie of Plantes*

ny blew or skie coloured flowers, growing for the most part vpon one side of the stalke. The roote is great, thicke and full of slimie iuice.

There is come vnto vs from beyond the seas diuers other sorts, whose figures are not extant with vs, of which there is one like vnto the first of these orientall Iacints, sauing that the flowers thereof are purple coloured.

Likewise there is another called *Orientalis albus*, differing also from the others in colour of the flowers, for that these are very white, and the others blew.

There is another called *Hyacinthus Brumalis*, or winter Iacint, it is like the others in shape, but differeth in the time of flowring. ✲ *The place.*

These kinds of Iacints haue been brought from beyond the seas, some out of one countrey and some out of others, especially from the East countries, whereof they tooke their names *Orientalis*.

✲ *The time.*

They flower from the end of Ianuary vnto the end of Aprill.

✲ *The names.*

There is a Lilly which *Ouid* in the tenth booke of his Metamorphosis called *Hyacinthus*, of the boy *Hyacinth*, of whose blood he faineth that this flower sprang, when he perished as he was playing with *Apollo*, for whose sake he saith that *Apollo* did print certaine letters and notes of his mourning writting thus,

Ecce cruor, qui fusus humo signauerat Herbas,
Desinit esse cruor, Tyrioque nitentior ostro
Flos oritur, formamque capit, quam Lilia, si non
Purpureus color his, argenteus esset in illis.
Non satis hoc Phœbo est, (is enim fuit auctor honoris)
Ipse suos gemitus foliis inscribit, & αι αι
Flos habet inscriptum, funestaque litera ducta est.

That is,

Behold the blood of him
Which dide the grasse, ceast blood to be, and vp there sprang a trim
And goodly flower, more orient then the purple cloth in graine:
In shape a lillie, were it not that lillies do remaine
Of siluer colour, where as those of purple hewe are seene:
Although that *Phœbus* had the cause of this great honor beene,
Yet thought he not that same inough, and therefore did he wright
His sighs vpon the leaues thereof: and so in colour bright
The flower hath αι writ thereon, which letters are of griefe.

Theocritus also hath made mention of this Hyacinth in *Bions* Epitaph in the 19. Eidyll,

Now Hyacinth, those letters thine tell them and do not passe,
And take vpon thy leaues αἶ αἶ, hei (mourning notes) alasse.

Likewise *Virgill* hath written heereof in the third Eclog of his Bucolicks,

Et me Phœbus *amat*, Phœbo *sua semper apud me*
Munera sunt, lauri & suaue rubens Hyacinthus.

And me *Phœbus* loues, *Phœbus* hath his gifts alwaies with mee
Trees Laurell, flowers Hyacinth so sweete and red to see.

In like maner also *Nemesianus* in his second Eclog of his Bucolicks:

Te sine me, misero mihi lilia nigra videntur
Pallentesque Rosæ, nec dulce rubens Hyacinthus:
At si tu venias, & candida lilia fient
Purpureæque Rosæ, & dulce rubens Hyacinthus.

Me without thee, white Lillies seeme all blacke poore man to me
And Roses pale, vnsweete the ruddie Hyacinth will be:
But if so be thou com'st, O then the Lillies shall be white
And Roses red, and sweete the ruddy *Hyacinth* in sight.

The Hyacinthes are said to be red, which *Ouid* calleth purple, some would haue them called by the name of Rustic Iron, for that when the best iron is made hot and wrought til it be cold, there remaineth a certaine blewnes, which they call iron colour: grounding themselues vpon *Virgils* auctoritie,

from this perspective the poetic anthologies that became popular in England in the second half of the sixteenth century—anthologies that, in their turn, frequently identify themselves, in their titles, through botanical metaphors.

Gerard goes on to discuss the known names of this particular kind of hyacinth; like most of the plants that he and other herbalists discuss, it has more than one, having been named repeatedly by its various discoverers and cultivators through history. In this case, Gerard notes that "[t]he first of our English Iacints is called *Hiacinthus Anglicus*, for that it is thought to grow more plentifully in England than elsewhere"; he adds that it is also called, by the Flemish herbalist Rembert Dodoens, "*Hyacinthus non scriptus*, or the vnwritten Iacint" (100). This brief passage elucidates a number of remarkable connections between early modern plants and texts at the fundamental level of the proper noun. There is the irony of the "first of our English Iacints" being named "*Hiacinthus Anglicus*," since such a name fails to distinguish this particular hyacinth from the other English hyacinths; its failure points to the tenuous state of botanical naming in Gerard's period. The plant's possession of more than one name, and Gerard's simple paratactic placement of the two names side-by-side, without any adjudication as to which should be propagated as the *proper* proper noun, also evokes something of the period's epistemology of naming and the desired or expected relationship between words and things. The fact that the plant's names are both given in Latin shows the connection of early modern vernacular botany to a classical inheritance and its contemporary humanist interpretation. Finally, the at first rather peculiar idea of an "vnwritten" plant may bring to mind Cordelia's description, in *King Lear*, of the "unpublished virtues of the earth," a phrase in which plants and writing are also joined; Shakespeare shows, as do so many other poets in the period, that the habit of linking these two otherwise apparently disparate materials and fields was clearly not confined to herbalists.

In calling this hyacinth "*non scriptus*," Dodoens was likely considering an ancient poetic account of another sort of hyacinth—one that could be characterized as "*scriptus*" instead. Gerard retells an account of this other hyacinth, taking his cue from Ovid (101–2). He quotes Ovid's verse at length and provides related excerpts from the idylls of Theocritus and from Virgil's eclogues, further entrenching his *Historie of Plants* in the history of poetry. More than half of the page that offers this part of the description of the hyacinth consists of block quotations of verse; on this and similar pages, the very typographical layout of the herbal comes to resemble an anthology of verse or a printed commonplace book (Figure 0.1). Prose and verse mingled in the herbals, as did fact and fiction: Gerard describes Ovid's tale as one that "faineth" (101) the origin of the hyacinth, yet happily incorporates this poetic fiction into his own predominantly factual account.

Not only does Gerard tell, in the midst of his botanical description, a poetic tale of a plant; in the story itself, writing and plants are mysteriously intertwined:

> There is a Lilly which *Ouid* in the tenth booke of his Metamorphosis called *Hyacinthus*, of the boy *Hyacinth*, of whose blood he faineth that this flower sprang, when he perished as he was playing with *Apollo*, for whose sake he saith that *Apollo* did print certaine letters and notes of his mourning[,] writing thus ...

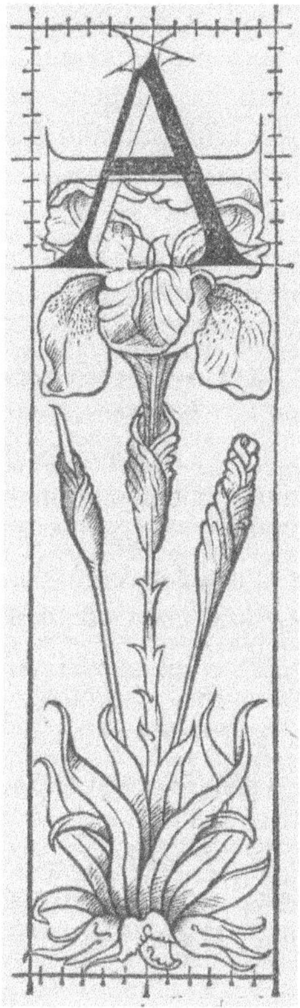

Fig. 0.2 An illustration from Geoffroy Tory's *Champ Fleury*

Here Gerard quotes Ovid and then translates him:

> Behold the blood of him
> Which dide the grasse, ceast blood to be, and vp there sprang a trim
> And goodly flower, more orient then the purple cloth in graine:
> In shape a lillie, were it not that lilies do remaine
> Of siluer colour, where as those of purple hewe are seene:
> Although that *Phoebus* had the cause of this great honor beene,
> Yet thought he not that same inough, and therefore did he wright
> His sighs vpon the leaues thereof: and so in colour bright
> The flower hath αι writ thereon, which letters are of griefe. (101)

The pattern in the petals of each and every hyacinth is thus read as reproducing in letters—like products of a primordial botanical printing press—an ancient expression of a god's grief. Gerard's anecdote demonstrates a sound classical precedent for the close contemporary association between plants and writing. He asserts the continued power of that precedent in his own culture when he concludes by noting that "most of the later herbarists do call this plant *Hyacinthus Poeticus*, or Poets Hyacinth" (102). The poets wrote meaning into the flower by telling the story of Apollo; the flower, in turn, has the poets' authorship inscribed in its own name. The recursive relations between plants and writing in this particular example highlight the intricacies of their imagined interpenetration.

Earlier in the sixteenth century, the French printer Geoffroy Tory wrote a book about the origins of writing in which he posited a related and even closer association between plants and written language. In *Champ Fleury* (1529)—not incidentally, a book about writing with a botanical metaphor embedded in its title—Tory suggests that the letters "I and A were conceived in the flower of a purple-hued lily called in Paris *lisflambe*" (75). As evidence, he includes an illustration with the letter A superimposed upon a *lisflambe* and enclosed in a square (Figure 0.2). He explains that the placement of the letter in the square was "represented by the word *Hyacinthus*, which consists of four syllables, *Hy-a-cin-thus*" (26). The very material form of letters is thus theorized to derive from the material form of plants, just as the syllables in a plant's name are read as if the structure of the word must necessarily correspond to the plant's shape. Verbal and herbal materials are integrated as closely as possible in Tory's account.

Tory, like Gerard, tells the story of Apollo accidentally killing his beloved Hyacinthus and atoning by turning the dead boy into a lily. Moreover, Apollo "made in the said lily two letters, Y and A, which we can still see there in some sort, traced in black and yellow on the petals" (24). At this point, Tory's text on the origin of the alphabet starts to read very much like a herbal in its discussion of the plant's various names, physiological qualities, practical uses, and consideration by prior herbalists:

> The whole plant is called by some doctors HIPIΣ, by others Gladiolus. The root has a sweet odour, and is mixed with other sweet-smelling things for keeping linen in chests. Marcellus Virgilius, Florentine scrivener, & commentator on Dioscorides, takes great pains, at Chapter LVIII of Book IIII of said Dioscorides, to give it to be understood that Hyacinthus is the lily which the Parrhisians call *lisflambe*; .... I have also heard and learned from the Florentines & and other well-informed Italians that the *lisflambe* is called in vulgar Italian Hyacinthol. For which reason it seems to me, under correction, that Hyacinthus and *lisflambe* are one & the same. (24)

Just as Gerard's history of plants starts to sound, in his account of the hyacinth, like a literary history, Tory's history of writing echoes the typical herbal in its decoding of botanical naming. Tory and Gerard's accounts merge further when Tory turns from his herbal commentary to quote the same Ovidian account and Virgilian references as Gerard. These two examples offer a glimpse of the extensive range of unexpected ways in which plants and texts were imagined in relation to each other in this period. This book explores those relations in their linguistic, conceptual, material and metaphorical forms. In particular, it concentrates on acts and examples of textual and botanical collection and the ways in which they intertwined in English culture in the sixteenth century.

Acknowledgments

Institutions that have supported my work include McMaster University, Queen's University, Concordia University, and Brock University, all publicly-supported Canadian universities; I am grateful for the opportunities for discovery and collegiality that they have provided me. The support of the Social Sciences and Humanities Research Council of Canada enabled the doctoral research upon which this book is based, while the Humanities Research Institute at Brock University has assisted with the final stages of its publication. Some of the material in this book was improved by thoughtful audience members at meetings of, among others, the Sixteenth-Century Society, the Group for Early Modern Cultural Studies, the Montreal British History Society, and the Renaissance Society of America. The eagle-eyed Holly McPhee helped enormously in editing the manuscript, as have many at Ashgate, especially Erika Gaffney and Seth Hibbert. I would especially like to express gratitude to my academic mentors during the formation of this work, among them Marta Straznicky, Robert Tittler and Kevin Pask, for the time and energy they have dedicated to me and my work. I am also grateful to Paul Yachnin and the many other good people of the Making Publics research project in Montreal. Jean Wilson, Elizabeth Hanson, George Logan, and Elizabeth Sauer have been particularly influential in wonderful ways. Many colleagues in the English department at Brock and beyond have also been superbly helpful. But without my friends and family, this book would not be. This is especially true of my son, who kindly waited until the manuscript was mailed off (just) to be born. To him, my parents, my brother, my Jim, especially: my love and thanks, always.

Chapter 1
Published Virtues of the Earth: An Introduction

A surprising number of early modern English books are identified in their titles as collections of plants. The English Short Title Catalogue turns up a plethora of printed works depicted as posies, nosegays, gardens, orchards, forests, sylvas, arbours, and bowers. In 1569, for instance, appeared the first of several anonymous editions of *A godlie gardeine, out of the which most comfortable herbes may be gathered for the health of the wounded conscience.*[1] Devotional books like this one frequently deployed botanical metaphors that emphasized their health-giving properties. More secular and more spectacular in its use of the same sort of titular trope is George Gascoigne's bouquet of 1573,

> *A Hundreth sundrie Flowres bounde vp in one small Poesie, Gathered partely (by translation) in the fyne outlandish Gardins of Euripedes, Ouid, Petrarke, Ariosto, and others; and partly by inuention, out of our owne fruitefull Orchardes in Englande: Yelding sundrie sweete sauours of Tragical, Comical, and Morall Discourses, bothe pleasant and profitable to the well smellyng noses of learned Readers.*

A similarly microcosmic botanical garden was envisioned in 1577 on the title page of John Bishop's

> *Beautifull blossomes, gathered ... from the best trees of all kyndes, diuine, philosophicall, astronomicall, cosmographical, historical, & humane, that are growing in Greece, Latium, and Arabia, and some also in vulgar orchards, as wel fro[m] those that in auncien time were grafted, as also from them which haue with skilful head and hand beene of late yeares, yea, and in our dayes planted: to the vnspeakable, both pleasure and profite of all such as wil vouchsafe to vse them.*

Around the same time, Richard Jones printed the more modest *Smale handfull of fragrant flowers, selected and gathered out of the louely garden of sacred scriptures, fit for any honorable or woorshipfull gentlewoman to smell vnto* (Breton); here the metaphor is conventionally gendered and moralized, but a couple of decades later Jones altered the same metaphor to emphasize the architectural features that

[1] The date on the title page (1659) is a printer's error, according to the English Short Title Catalogue record.

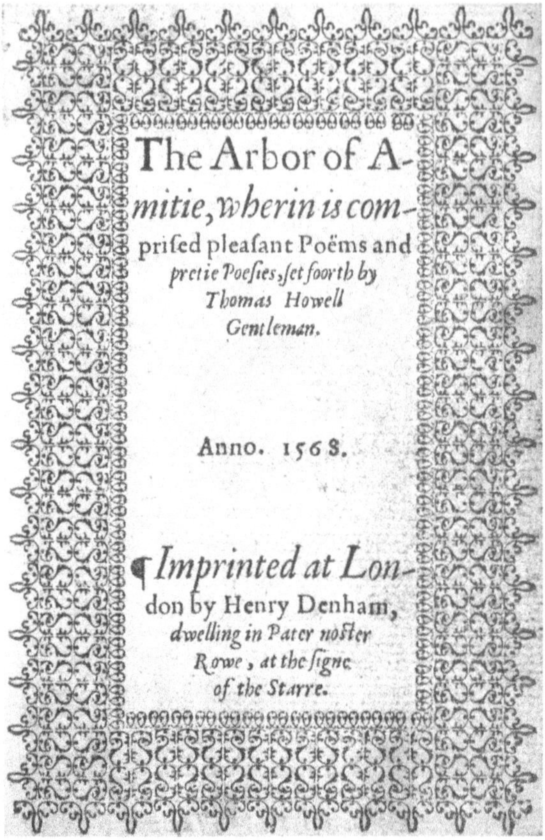
The Arbor of A-
mitie, wherin is com-
priſed pleaſant Poëms and
pretie Poeſies, ſet foorth by
Thomas Howell
Gentleman.

Anno. 1568.

¶ *Imprinted at Lon-*
don by Henry Denham,
dwelling in Pater noster
Rowe, at the ſigne
of the Starre.

Fig. 1.1 The title page of Thomas Howell's *The Arbor of Amitie*

framed contemporary gardens in two more titillating titles: *Brittons bowre of delights* (Breton) and *The arbor of amorous deuises* (Breton).

These titular choices had a sound linguistic foundation, since a number of English words for textual collection—the florilegium, the sylva, and the anthology—allude etymologically to plants. Moreover, many such books contain collections of brief lyric or epigrammatic poems as well as sententious quotations and proverbs of the kind then sometimes known individually as posies—a term easily conflated in its singular form with poesy, a period term for all imaginative writing. Even some apparently non-vegetable terms for textual collection used in the period may be traced to a botanical root. Jack Goody notes that the term *peristephanon* ("crown"), a title used by the early Christian poet Prudentius, referred to imperial Rome's honorary floral coronas (66, 106); early modern coronas of sonnets, like those of John Donne and Mary Wroth, may thus be further if more obscure instances of the botanical tropes repeatedly chosen by early modern writers and printers to represent the act of textual collection.

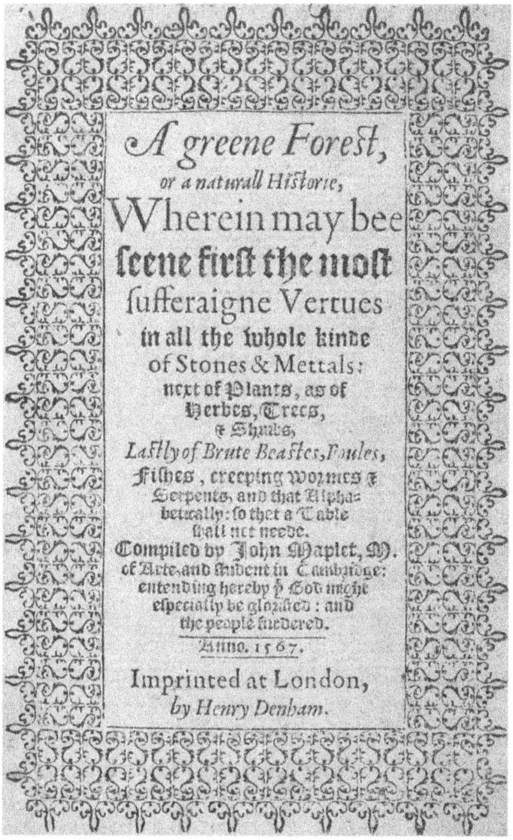

A greene Forest,
or a naturall Historie,
Wherein may bee
seene first the most
sufferaigne Vertues
in all the whole kinde
of Stones & Mettals:
next of Plants, as of
Herbes, Trees,
& Shrubs,
Lastly of Brute Beastes, Foules,
Fishes, creeping wormes &
Serpents, and that Alpha-
betically: so that a Table
shall not neede.
Compiled by John Maplet, M.
of Arte, and student in Cambridge:
entending hereby y God might
especially be glorified: and
the people furdered.
Anno. 1567.
Imprinted at London,
by Henry Denham.

Fig. 1.2 The title page of John Maplet's *A Greene Forest*

Most books masquerading in these ways as gardens and garlands were not remotely botanical in terms of the material they treated, but plant metaphors occasionally also appeared in the titles of texts that really did take plants as their subject matter, such as the pioneering printed book of herbal medicine known variously as the *Hortus* or the *Ortus Sanitatis*: the garden of health. The trope could thus be used in more and less metaphorical degrees. Moreover, the same author or printer might appeal to a botanical trope in utterly different genres: Hugh Plat's *Floures of philosophie* and his *Floraes paradise*, for instance, are a collection of Latin sententiae and a horticultural work, respectively. Similarly, when Henry Denham printed *A greene forest* (Maplet) in 1567 and *The arbor of amitie* (Howell) the very next year, the botanical metaphors conflated instead of distinguishing their different contents: the first is a natural history and the latter a collection of poetry. Their title pages, boxed in by similar knotty hedges of printer's fleurons (Figures 1.1 and 1.2), also fail to disclose the generic difference at first sight. A twenty-first century bookstore browser, on the other hand, rarely mistakes a biology textbook for a slim volume of verse.

Early modern elisions between books about gardens and books pretending to *be* gardens extended beyond first impressions, since the prefatory rhetoric found in horticultural handbooks sometimes sounds as if it had been lifted from a volume of verse. One writer addresses his readers thus: "To satisfie therefore their desires that are louers of such Delights, I took vpon me this labour and charge, and haue here selected and set forth a Garden of all the chiefest for choyce, and fairest for shewe." Another author, writing in a very different genre, instructs his readers to "[m]arke then, what varietie of flowres grow all along as thou goest, and trample on none rudely, for all are right precious. If thy conscience be wounded, here are store of hearbs to heale it .... In briefe, what infirmitie canst thou have, but here it may bee cured?" The first passage comes from John Parkinson's epistle to the reader in his *Paradisi in Sole*, a horticultural handbook, while the second appears in John Bodenham's *Bel-vedére or, The garden of the muses* (A3v), a verse anthology—but the passages themselves provide no hint of the distinction in kind. Indeed, the aesthetic emphasis of the first passage, on "[d]elights" and fair shows, might seem more fitting in a book of poetry, while the attention in the second to botanical healing powers sounds more appropriate for a medicinal herbal.

As these examples clarify, plants and texts both belonged to a broader culture of collecting in the period. Plants and texts were both collectible objects, susceptible to encyclopedic or selective gathering, while gardens and books were sites in which to store and display these objects and others. It is most often compilatory texts that are portrayed as gatherings of plants; likewise, it was the activities surrounding plant-collecting that enabled or required the textual counterparts of herbal encyclopedias and herbaria. Lisa Jardine notes the likeness of the extraordinarily rapid and reliable reproductive processes available both to patrons of the press and of bedding plants: "[i]n this respect the book trade, by the 1550s, closely resembled the trade in tulip bulbs at the same date" (165). Collectors imagined and treated these collectibles similarly, she suggests, because they were similarly amenable to division and replication. The same was not generally true of other collectibles, whether natural or artificial. Mammals, fish, and birds portrayed in early modern cabinets generally appear emphatically infertile: dried-out singularities fastened to the ceiling or arranged statically on shelves. Often only parts of a specimen's carcass would last, so that beaks, claws, and bones had to stand in, synechdochically, for lost flesh. Artificialia, such as coins and medals, were similarly singular except in the reduced form of rubbings and illustrations. Both texts and plants, on the other hand, could be multiplied and shared with relative ease, and without detriment to the original collector. Yet this unusual material quality was just one of the ways, I suggest, in which plants and texts occupied similarly distinct places in an early modern culture of collection.

Yet such intersections between botanical and textual collections have received little sustained study. Significant interdisciplinary studies have, however, established a central role for collecting in early modern culture. Amy Boesky insists that "[t]he collection and organization of objects in the period was not merely a metaphor for cultural practice ... but an entire epistemological construction" (313; also see Mullaney). More particularly, Oliver Impey and Arthur MacGregor note in their collection on cabinets of curiosities that "[i]nterest in the natural world was a major preoccupation of Renaissance learning, and here collecting was to play an indispensable role" (1). Paula Findlen's *Possessing Nature: Museums, Collecting, and Scientific Culture in Early Modern Italy* is a remarkable study of such sixteenth-century natural-history collections in a single cultural context. While there is no similar monograph on English natural-historical collections of the same period, Marjorie Swann's *Curiosities and Texts: The Culture of Collecting in Early Modern England* explores and characterizes a widespread variety of similar collections in the seventeenth century. Material and textual versions of collections are closely integrated in her study, but the literary collections most closely identified with the trope of the garden of verse have generally been treated as phenomena apart from the contemporary collections of material, and particularly natural, artifacts.

The nature of early modern habits of textual collection have, however, received substantial attention in themselves, and in the course of such studies the prevalence of botanical metaphors has also been treated. Terence Cave's analysis of "the double figure of *copia* and cornucopia as a thread which may be followed through the labyrinth of sixteenth-century writing" (xv) provides a broad intellectual history of the period's use of botanical tropes in textual collections, as does Ann Moss's tracking of the classical and medieval origins of the trope of the scholarly humanist bee, diligently gathering nectars from far-flung textual flowers and compounding them into a honey of their own making. Crane characterizes the self-conscious practises—including botanical tropes—shaping textual collections in sixteenth-century England, while as part of a broader study Bushnell explores the prominence of related botanical metaphors applied to humanist pedagogical practices, noting that "[t]he most frequently cited metaphor for the book in early humanist pedagogy was that of the garden" (*A Culture of Teaching* 135). The most direct study to date of such titular botanical metaphors in English print culture may be found in an essay by Randall L. Anderson, which appends a comprehensive inventory of relevant early modern publications. Despite such extensive interest, both in the period's collections of *naturalia* and in textual collections governed by botanical tropes, gardens of verse have rarely been set in close relation to the contemporary culture of botanical cultivation and collection. That is, material and metaphorical plants have not generally been examined or understood in conjunction, as they so often were in the early modern period, and as they will be here.

This is a study of the plants that grew and some of the texts that grew up around them in late sixteenth-century England, including not only the more familiar gardens of verse but the herbal encyclopedias and herbaria that played so important a part in the botanical culture of the day. While printed herbals have been extensively studied, comparatively little has been said about the writing practices in herbals in relation to the larger literary culture in which these books took form.[2] Those who

[2] Agnes Arber's authoritative if somewhat inevitably dated *Herbals*, first published in the early twentieth century by a distinguished botanist, does include some analysis of the nature of descriptive writing in the early herbals, while Brian Ogilvie's *The Science of Describing* treats the subject expansively in a larger study of the development of

have addressed these botanical texts have usually done so from the perspectives of historians of science or of the visual arts.[3] Consequently, they rarely foreground the linguistic, literary, textual, rhetorical, or more broadly bookish matters raised by their primary sources. It is to these matters that I would like to direct my attention.

When she observes her father's imperiled health, Cordelia cries out for help from the "vnpublisht vertues of the earth" (Shakespeare, *M. William Shak-speare* 1v)—but by the time her words were published in 1608, a very long list of medicinally-virtuous plants in England had secured publication. The second half of the sixteenth century corresponds with an English botanical renaissance that kept pace with the much more familiar literary renaissance of the same era. It was in these decades that the physician William Turner and the surgeon John Gerard published their vernacular herbal work; simultaneously, the innovations of the botanical garden and the herbarium had their first impact. English plant culture, ranging from such medicinal herbalism to the more familiar practices of recreational gardening, was correspondingly transformed in ways that may be measured, in part, by the textual production that burgeoned to match the increased interest in these fields. One bibliographer of botanical publications points to a bookish bumper crop, beginning around the middle of the century and peaking at its end (Henrey 3). The historian of botany Julius von Sachs notes that while the invention of botanical gardens "show how lively an interest was taken in botany in the latter half of the sixteenth century[,] this is still more shown by the great number of books of plants" (19). Productivity in one field was paralleled by productivity in the other.

By mid-century, London was populated by a set of botanical enthusiasts who kept not just recreational but experimental gardens (McLean 220–21) and whose novel spadework was newly supported by the earliest English handbook on gardening, Thomas Hill's *How to dresse, sowe, and set a garden*, which was in turn followed by "a steady stream of gardening literature ... from the 1550s onwards" (Cantor 111).

European natural historical studies and writing from the fifteenth through the seventeenth centuries. For more incidental treatment, see Eleanour S. Rohde's *Old English Herbals*, an early bibliographic survey; Blanche Henrey offers a more recent and detailed history and bibliography.

[3] A recent and magnificent exception is Anna Pavord's comprehensive and interdisciplinary *The Naming of Names: The Search for Order in the World of Plants*. Herbals are considered (although also often dismissed as merely pre-scientific) in many histories of botany by, among others, J. Reynolds Green, R.T. Gunther, A.G. Morton, Julius von Sachs, and Edward Lee Greene. Among older histories of natural history, Charles Raven's *English Naturalists from Neckam to Ray* is outstanding, and includes insightful chapters on William Turner and John Gerard, as well as a broader analysis of the development of English natural history. Brian Ogilvie's *The Science of Describing* provides a recent and thorough examination of the developments in European Renaissance natural history. For herbals in art history, see, among others, Wilfrid Blunt's *The Art of Botanical Illustration*, Blunt and Sandra Raphael's *The Illustrated Herbal*, Vera Kaden's *The Illustration of Plants and Gardens 1500–1850*, Lys De Bray's *The Art of Botanical Illustration: The Classic Illustrators and Their Achievements from 1550 to 1900*, and Gill Saunders's *Picturing Plants: An Analytical History of Botanical Illustration*.

One historian of natural history has traced the growing interest in gardens by following the changes between the editions of William Harrison's *Description of England* of 1577 and 1587, with the second discovered to be "much amplified in its references to nature" and to contain an added chapter on gardens and orchards (Raven 193). Surviving maps confirm that in the latter half of the century, enclosed gardens were found both inside and outside of London's walls (Schofield 314). Meanwhile, still further outside those walls, new crops began to be cultivated and both old and new agricultural manuals also became marketable commodities (Bushnell, *Culture* 86; Morton 151). Thomas Tusser's *Hundreth good pointes of husbandrie* of 1557, for instance, was reprinted and expanded many times before the end of the century. In a kind of symbiosis that crossed the boundary between art and nature, developments in plant culture played out in textual culture.

This was a time when people interacted with and depended on plants in ways now generally forgotten or concealed by industrialization. As Thomas Johnson wrote in his 1633 edition of Gerard's herbal: "God of his infinit goodnesse and bountie hath by the *medium* of Plants, bestowed almost all food, clothing, and medicine vpon man" (2). Something as essential as pharmacy was largely and conspicuously based on plant material (Siraisi 148), and something as frivolous and frequently condemned as cosmetics was as well (Eyler 26). Few today make their own ink from insect-inhabited galls growing on oak trees (Finlay 26–7). Households that could afford the luxury were strewn with enough rushes, fragrant flowers, and herbs to have the habit remarked upon by visitors from abroad (Peter Thornton 14). Even London itself was not the concrete jungle of today, with the countryside only two miles from the city's centre, where verdure flourished anyhow in both household and market gardens (Amherst 101). The success of local harvests was both more visible and more critical than in a globalized market, since when they failed, people rapidly went hungry (Appleby 105–14). Timber stands, the source of both fuel and building material, were also often at least perceived to be in short supply (Cantor 101; Henrey 101–2). In the midst of perceived and actual dearth, however, a burgeoning variety of vegetables and fruits were hawked on the streets and welcomed to tables as a cornucopia of previously unheard of plants arrived from the Americas and other parts of the world. In less than a century of overseas voyages, over twenty times as many new plants arrived in Europe as had appeared in the previous two millennia (Morton 118). The effect may be observed in the English print record, in which the minuscule number of plants in Turner's first book (134) and the considerably larger but still small number in his second book (almost 400) were dwarfed not a century later when John Parkinson's *Theatrum botanicum* would include descriptions of almost 4,000 plants. Of all these novelties, those with medicinal uses were most valuable and—along with spices, another botanical treasure—rated second only to precious metals (Morton 119). Foreign finds in turn inspired new explorations and applications of native and local flora. Whether flourishing or failing, such varied manifestations of plant culture were not easy either to take for granted or to ignore.

A more strictly academic botany began to emerge from conventional medicinal herbalism when European professorships in simples were instituted in the 1540s. But botanical natural history had long been a bookish pursuit. It was not predominantly an experimental but a historical science, relying more on the authority of ancient texts than on personal observation or experiment. The increasingly elaborate nature of the herbals printed throughout the sixteenth century, however, can largely be ascribed to the relatively novel (at least in England) humanist project to authenticate this classical textual inheritance. The humanist revivification of classical texts encouraged quite literally ground-breaking work in the non-therapeutic study of plants, one of the earliest but today least popularly regarded manifestations of modern science. For instance, while William Turner's first botanical books concentrated on cross-referencing classical and vernacular botanical names, Turner is equally famous for his far-flung field trips—for digging in the dirt as well as in books. His example demonstrates how a humanist interest in ancient botany merged in this period with a more popular interest in home-grown herbalism, one that made a transition at this time from a classical to a vernacular focus.

All of these books on plants—agricultural, horticultural, and pharmaceutical—were produced when the printed book and literacy more broadly began to thrive in England.[4] The formation of the Stationer's Company in 1557 points to the definitive arrival of print as a cultural industry; in the same year, Richard Tottel published both his best-selling and pioneering anthology of poetry, *Songes and sonnettes*, and Tusser's husbandry manual—the latter also, not incidentally, in verse form. The genre of the printed poetic anthology, which is perhaps the most familiar and typical species of textual garden, reached a zenith of popularity and production after Tottel that increased up to around the year 1600 (Pomeroy 29–30); similarly, it was during the 1590s that English botanical and horticultural literature was most prolific (Henrey 3). One scholar suggests that the taste for textual collections—in which writings were often harvested from one context and grafted into others—in fact crossed over at this time from poetry to natural history: William Ashworth argues that Erasmus's frequently printed anthology of adages "helped create a taste for the clever, pithy aphorism that, by the middle of the sixteenth century, spread to include observations about the natural world" ("Natural History" 310). The fifty years that followed grew unusually rich in such botanical and bibliographic interactions.

Even in their sheer raw materials, plants and books were remarkably intertwined in this period. Books depended increasingly for their material existence not on livestock, for vellum or parchment, but on crops, since paper at this time came not from softwood but from flax, via recycled linen rags. Their common material origin could also determine a similar fate for neglected fruits and books, as Spenser

[4] Tessa Watt discusses the years from 1560 to 1580 as a period in which all social groups markedly improved their literacy levels, with the percentage of fully literate husbandmen improving from 10 to 30% and literate yeomen increasing from 45 to 75%, for instance (260). See also her summary of the complexities of evaluating literacy in this period, when many more people may have been have been readers than writers (6).

shows in his portrait of a library "all worm-eaten, and full of canker holes" (*Faerie Queene*, 2.9.57, line 9). Yet the means to preserve books from worms was also to be found in plants: Albertus Magnus notes in his *De vegetabilibus* that wormwood was so named for its ability to repel worms and mice from books, while other plants were put to similar uses in libraries (Perry 137–9). Book chests often contained both plants and books in close proximity; one historian of furniture notes that early English chests often have a "little ledge or shelf just under the lid at the side .... to accommodate lavender or some other sweet herb" (Gloag 38–9), the effect of which would be not merely fragrant but preservative, as is suggested in the book known as Banckes' herbal, where rosemary is advised for just such a purpose: "take the flouers and put them in a cheste amonge your clothes, or amonge bokes and moughtes [moths], shall not hurt the[m]" (Copland).

Books depended on plants for their material composition and preservation in still other ways. Covers were made from wooden boards (Lisa Jardine 141), or sometimes leather, which required tree bark in its tanning, as Gerard notes in his herbal (35). Both Gerard and Turner (in the 1562 volume of his herbal) note that one flower's gluey juices could serve as a paste to fasten pages together into home-made booklets (Gerard, *Herball* 99; Turner 398). Most illustrations in sixteenth-century English books were woodcuts; Gerard notes the special suitability of pear wood, which "serueth to be cut into many kindes of mouldes" including "such prints as these figures are made of" (*Herball* 1271). Writing ink, as well as home-made ink for colouring pictures, often derived from galls growing on trees and other botanical materials (Plant 186; Finlay 26–7). Gerard describes one such recipe: "if you lay sap berries in steepe in faire water for the space of two houres, and mixe a little Saffron with that infusion, and laie it vpon paper, it sheweth the perfect colour to limne, or illumine the flower withall" (*Herball* 153); in other words, flowers were used to paint flowers, neatly closing the gap between the artistic medium and its message. Some of the early modern pens that made use of such botanical inks were themselves made out of reeds (McKitterick 15), as were some of the candles by the light of which one read (Thomas 73). Sand was scattered on paper to blot hand-written works, temporarily disguising the page as a tiny plot of earth. Printing introduced still more material and metaphorical links between plants and books. One clever way to meet the higher demand for paper connected print culture to agriculture when Dutch papermakers, in the late sixteenth or early seventeenth century, altered a seed-grinding machine so that it could pulp rags instead (Avrin 295). Printer's ink was made partly of varnish derived from boiled linseed oil (Plant 186). The generic illustrative ornaments used by printers were called fleurons and tend to resemble above all else bird's-eye views of the herbaceous borders of the typical Tudor knot garden. The wooden printing press itself may have been modelled on, and at any rate resembled in its manufacture and method, presses used on grapes, olives, oilseeds, and herbs (Moran 19). The entire process of printing bore resemblance to the activities of the agricultural calendar, from the initial sowing of the seed-like type, through the plowing of the page with an imprint of those letters, to the seemingly endless harvest of the hundreds of sheets that proliferated from the press, which in turn needed to be gathered to dry, like grain in a grange, before they could be assembled for market. Printing also, as mentioned, endowed writing with a reproductive system that, in its profusion, speed, and high fidelity, closely resembled the otherwise uniquely prolific fertility of plants.

The material connections between plants and books were complemented by a host of linguistic conjunctions. Botanical titles demonstrate how often the language used to discuss books, texts, and writing appealed to an imaginative link between the verbal and the herbal. This was as true of classical languages as of English: the Greek ancestor of "bible" referred to the inner bark of the papyrus of which it was made, the Latin "liber" to a tree's bark and "codex" to its trunk, while "book" itself may derive from a Germanic word for the beech. "Cultus" had long signified both agricultural tillage and the ploughing of the cloddish mind through a literary education aimed at making it more fertile; similarly, "versus" could signify either a line of verse or a furrow (Herron 101). In sixteenth-century English, the term "tong" was equally applicable to the speaking tongue and the head of a plough, a pun all the more appropriate in an age when so much verse was intended for recitation. Spenser appears to use the pun when he asks his muse, in the proem to the first book of *The Faerie Queene*, to "sharpen my dull tong" (2.9). That the term could be used not only in reference to personal diction but to the language as a literary tool is suggested by the prefatory lament, in a 1567 translation of Xenophon, that "our grosse tongue is a rude and a barren tong, when it is compared with so florishinge and plentifull a tongue" as Greek. The word "slips" could refer either to pieces of paper or to cuttings from plants, as Isabella Whitney shows in her instructions about how best to employ her *Sweet Nosgay*: "when you come into a pestilent aire that might infect your sound minde: yet savour to these SLIPS in which I trust you shal finde safety" (Av^{r}); the double entendre linking books and plants is emphasized typographically, perhaps indicating the special pleasure such paronomasia engendered in the Elizabethan imagination. Similarly, words like "leaves" and "sheaves," then as now, referred equally to pages and to foliage—as, of course, did "folio." Other homonyms linking the fields of vegetable and textual production are now more obscure; for instance, grafting was of considerable importance in English fruit farming (Henrey 55), but the word at the time was generally spelled "graff," the etymology of which derived from the Latin and Greek words for writing and connected to the Old French word for stylus or pencil.[5] More historically-specific instances may also be cited; resemblances between certain material qualities of paper and plants may have been remarked upon metaphorically, for instance, in the designation of

---

5 The *Oxford English Dictionary* has "graff," being "superseded in ordinary use by graft" and deriving from "late L. graphium" and "Gr. grafi'on, grafeion stylus, f. gra'fein to write. The sense 'stylus, pencil' is common in OFr.; the transferred sense of 'scion, graft' was suggested by the similarity of shape." A browse through a recent gardening catalogue shows that today's amateur gardeners are advised to use a common pencil sharpener to fashion their grafts.

certain types of tulips as *ghemarmerde, marbrée,* or marbled, likely in reference to the marbled paper that reached Europe from Turkey concurrently with the tulip and that was a similarly valuable collectible object (Goldgar 332). Even abstract literary terms could forge connections between plants and texts: for instance, the word "plot," toward the end of the sixteenth century, "pervaded popular discourse in new and interconnected ways: the explosion of surveying manuals (guidebooks for charting the land) corresponded with the development of the idea of narrative plot (literal and figurative charts of a story line)" (Brückner and Poole 618). It may well have been hard to speak about words or books without reference to the vegetable kingdom, and vice versa.

The material facts and metaphorical fictions about the relations among plants and books are thus not easily separated; nor is such a separation likely to clarify how those relations were imagined in the early modern period. In the following chapters I examine some of those relations as they were perceived and portrayed, both in the garden and on the page, during a pan-European botanical boom that cropped up precisely when textual collections of vernacular flowers became a factor on the English book market. The next chapter explores the humanistic mode of plant culture, cultivation, and collection as it took on new form in the middle of the sixteenth century, when medicinal herbalism yielded the novel and interdependent botanical and bibliographic phenomena of experimental gardens, herbaria, and printed herbals. Chapters 3 and 4 each focus on a single English herbalist, closely reading their herbal writings as case studies of the particular connections between plant culture and print culture. I argue, for instance, that the Protestant doctor and divine William Turner was driven by the Reformation's valorization of vernacular translation and printing, while John Gerard's problematic involvement in the creation—or, some would say, plagiarism—of his herbal was bound up in the anthological and commonplacing literary culture that surrounded him. All three chapters highlight the intricacy in early modern botanical work of the apparently simple act of naming—an act in which words and things, texts and artifacts, are ideally (but rarely really) set in an elegant one-to-one relation. The last chapter returns to the textual collections of verse and aphorism disguised as gardens and groves that were printed so prolifically in the Elizabethan period in order to read their governing metaphors as marketing tools that helped transplant books into a domestic setting: the established household place of plants could provide a model for the integration of the relatively novel artifact of the printed book, and texts that likened themselves to plant material made themselves more desirable by purporting to be similarly useful and beautiful things to be collected and preserved against need. As a whole, this study demonstrates the range of ways in which plants and texts occupied remarkably similar and interdependent places in a larger culture of collection in sixteenth-century England, as is evidenced by their linguistic, conceptual, metaphorical, and even material intersections. The still-familiar metaphor of the "garden of verse" provided the people of this period with an imaginative model for the collection, display, and reproduction of texts at a time when these concerns were central to a literary culture that was being transformed by the technology of print. At the same time, the collecting of actual plants, and the books made to contain and represent those botanical collections, were influenced in sometimes unexpected ways by broader literary conventions of textual collection. In the fields of botany and books, one material was knowable in terms of the other because one was so often and so thoroughly embedded in the other. The following chapters aim to illuminate the resulting interplay among materials and discourses rarely considered in tandem today.

Chapter 2
The Bookish Nature of Botanical Culture: Continental Contexts

> [O]ur view of natural history has been distorted because we have not been asking ourselves the right questions .... Why did Renaissance scholars gather and publish information about the natural world? What kind of material was included in their compilations, and why? What was the intended audience of the publications; what was the intended use of the information contained in them?
>
> —William B. Ashworth, "Natural History and the Emblematic World View," 305

> The documents of this new history are not other words, texts or records, but unencumbered spaces in which things are juxtaposed: herbariums, collections, gardens; the locus of this history is a non-temporal rectangle in which, stripped of all commentary, of all enveloping language, creatures present themselves one beside another ...
>
> —Michel Foucault, *The Order of Things*, 131

I set these passages beside one another to suggest a comparison of two methods for investigating the history of natural history. Ashworth suggests that by asking better questions of the texts associated with Renaissance natural history, we can usefully revise our understanding of the subject. The passage from Foucault suggests a contrasting approach: rather than turning to the "words, texts or records" of natural history, we are to turn instead to the non-textual "spaces," like gardens, in which its objects and creatures can be found. Foucault's suggestion is, however, not prescriptive but descriptive. More importantly, he is not, like Ashworth, describing English "Renaissance" natural history, but that of the Baroque period, of what he calls the "Classical" episteme of the seventeenth century, which, he says, decisively breaks with the prior episteme of resemblances. This same break is said to have broken the bonds between *res* and *verba*, the things and words so consistently mingled in earlier approaches to what was called the Book of Nature. Spaces and objects, not texts, Foucault says, became the sites of a new history-free ("non-temporal") natural history, which in turn became the basis of subsequent scientific knowledge. These spaces and objects were, in the ideal of the period as he understands it, "unencumbered" and "stripped" not only of time but of text.

What if, instead, someone writing the history of natural history should consider the spaces Foucault names—"herbariums, collections, gardens"—in relation to a time when they were not yet "stripped of all commentary, of all enveloping language," by those who used them to contain and understand nature? Such an approach could answer Ashworth's call to consult the texts of natural history, and might also recognize just how expansive the period's sense of textuality was: how words were understood to interpenetrate with the very objects, spaces, and creatures that were later purified of the alleged taint of language. With such a hybrid approach in mind, I will examine in this chapter various textualized spaces in which medical botany developed in the sixteenth century—the experimental gardens, *horti sicci*, and herbals—to show how plants and books met in the emerging early modern culture of collecting. While the primary act of botanizing at this time may have been collecting plants, that act was twinned with another no less important one: re-collecting such a material collection in textual form. Frequently, this meant collecting plant names, the linguistic material most closely associated with botanical material, the closest *verba* gets to *res*. Unique and specialized spaces for the collection of plants depended, in turn, upon corresponding collections of texts and books.

Yet the progress made by the century's botanical pioneers is often attributed to their tendency to turn toward an original and personal observation of actual plants and their collection, and away from attention to traditional textual and pictorial representations that are often characterized as distorted or unrealistic. In many cases, however, advances in botany depended just as much upon a return to and a new recruitment of the representational forms, qualities, and capacities of the book, both pictorial and textual. While medieval natural philosophers may have emphasized the authority of literary tradition at the expense of material nature,[1] sixteenth-century naturalists did not, in their determination to understand the natural world, simply reverse these priorities by rejecting textual authority and the representational powers of the book. Instead, new habits of botanical collection tended to yield various textual counterparts, often in novel forms, that broadened the conventional associations between the material and metaphorical natures of texts and plants.

Agnes Arber, a botanist and early scholar of the oldest printed herbals, acknowledges the interrelation between botany and the book in this period: "[s]econd only to the debt of botany to medicine," she writes, "is its debt to certain branches of the arts, especially printing and wood-engraving" (*Herbals* 265). William T. Stearn, a former librarian of the Royal Horticultural Society, confirms the dependence of sixteenth-century plant culture on print culture: "Botany developed largely with the aid of printed works facilitating the international spread of local knowledge" ("Introduction" xxvi). More recently Paula Findlen has argued that sixteenth-century botanical natural history was shaped

[1] For instance, even in the relatively advanced culture of natural history that grew up in the twelfth century around such figures as Albertus Magnus, the "plants which inhabited nature were encountered, not in field and forest, but on the written page, for medieval intellectual culture was pre-eminently a culture of the book" (Peter Harrison 65), while from the thirteenth century on, "[t]he turn to nature as an entity in its own right was a turn to texts about nature" (67).

> *through books*: the visibility and endurance of natural history during the Renaissance was due in no small part to the success of its publications and to the discussions that arose because of them as well as to the copious collections of natural objects, near and far, that demanded closer scrutiny in order to be known. ("Formation" 372; emphasis hers)

Findlen emphasizes, as I do, the interplay between plants and books—artifacts and their representations—as the salient characteristic of sixteenth-century *res herbaria*. This chapter outlines the relations between a broadly European botanical culture and the changing bibliographical and textual cultures of the day, since continental work in the first half of the sixteenth century provided the basis for the developments in early modern English plant culture in the second. I begin by establishing the novel nature of the period's approaches to medicinal herbalism and then highlight some specific ways in which advances in the understanding of plants were intertwined with texts, books, and print culture. It will become apparent that early modern botanists relied on the accumulation and circulation of texts at least as much as of seeds and slips to establish an accurate and meaningful correspondence between these two very different materials.

Botanical Renaissance: A Potted History

Physicians early in the sixteenth century often had little contact with the plants that dominated the ingredients in their pharmacopeia. Instead, they generally understood plants primarily through the dense mediation of much copied ancient and medieval herbals. Many held the apothecaries who maintained collections of herbal simples in considerable contempt (Reeds 26). Apothecaries, after all, often depended in their turn on networks of untutored practitioners, commoners who supplied the pharmaceutical inventory by undertaking the dirty work of collecting plants in the first place (Frank Anderson 128, 137; Reeds 24). Physicians were not required to soil their own hands, with so many other and abler ones at work in traditional medicine, from the local clergymen to the women (often referred to as "herbwives") who kept collections of medicinal recipes and gardens. Illiteracy was increasingly cited by physicians as a justification for preventing apothecaries and women from practising as herbalists: "as male doctors increasingly organized their profession in the early modern period, terms such as *illiterati mulieres*, *indocti*, and *vulgares* came into more frequent use" (Weber 369). The growing focus on literacy as a credential in herbalism indicates an increasing attention to and valuation of the relations between plants and books. Even William Turner, a physician who showed himself willing to dig in the dirt to learn about the plants he depended on, identified his own instructor in herbalism—"Lucas Gynus [Ghini] / the reder of Dioscorides in Bonony [Bologna]" (1568 iii)—primarily in terms of his relation to the authoritative text in his field rather than to the plants with which he worked. One result of the division of labour between plant collectors and the physicians was that those who, with their extensive formal education, might have known the most about plants were instead often the least acquainted with them in their unprocessed form, since the distasteful manual labours of collection and decoction were left to underlings.

During the sixteenth century, however, some of these relationships among people and plants began to change. Physicians were subject to critique for being unable to name or recognize the botanical simples they prescribed in compound form. In his *Anatomy of Melancholy*, Robert Burton quotes the sixteenth-century herbalist and physician Leonhart Fuchs as having written that "[m]any an old wife or country woman doth often more good with a few known and common garden herbs than our bombast physicians with all their prodigious, sumptuous, far-fetched rare conjectural medicine" (213).[2] On the continent, botanical fieldwork for the purpose of plant identification and collection, once beneath the dignity of gentlemen students, became validated through the establishment of professorships in simples, which brought about concomitant changes in the medical curriculum (Hoeniger, "Plants and Animals" 145). Gradually, academic respect grew for an active study of plants alive in the fields in place of poring over time-honoured volumes. The emergence of academic botany is thus an early, earthy sign of scientific evolution that is often neglected in favour of more tumultuous cosmological revolutions.

At the same time, ambitious botanical gardens were instituted at continental universities just before the middle of the century. These were novel technologies, invented spaces in which to store, study, and perpetuate collected specimens. Such gardens provided spectacular signs of the academic study of the Book of Nature. Moreover, their presence on campus and in the curriculum meant that actual plant material was increasingly handled, collected, cultivated and understood by the medical elite, which enhanced the broader status of plants as commodities and as forms of cultural capital. Consequently, while botany may have begun as an outpost of medical faculties, it quickly attracted noble patronage, establishing its prestige in a larger cultural arena (Findlen, "Formation" 378). By the middle of the century, Italian princes and patricians were patronizing natural-historical research through the establishment of private botanical gardens (Swann 58). Humanist natural historians became highly sought after by the aristocracy.[3] Such continental trends made their way without too much delay to England. As early as the 1540s, the Duke of Somerset, then Lord Protector to King Edward VI, patronized William Turner at an estate notable for its rare specimens; a few decades on, Elizabeth's

2 Similar, and more notorious, is the story told by Erasmus of the physician Guillelmus Copus, who picked a leaf out of his salad at a dinner with members of the Parisian faculty of medicine, none of whom could name it and all of whom agreed it must be something quite exotic; Copus then called in a kitchen maid, who was able to identify it as parsley (Reeds 25).

3 Leonhart Fuchs, for instance, was a court physician. Andrea Cesalpino turned from professing botany at the University of Bologna to become the pope's physician. Pier Andrea Mattioli became physician to two Holy Roman Emperors (Findlen, "Formation" 377–8); he was succeeded by Rembert Dodoens and Carolus Clusius, while Matthias L'Obel was the physician of William the Silent and later the official botanist to James I (Brenninkmeijer-de Rooij 367; Greene 882).

trusted advisor William Cecil employed the surgeon and herbalist John Gerard as the chief gardener of his extensive properties.

While medical students and others may have been newly encouraged to make field trips and plant botanical gardens, books were not consequently rejected from the study of plants; rather, the changing relations among people and plants were mediated by the vigorous exploitation of a bookish culture that was undergoing its own remarkable developments. For one, a new factor was at play in the already considerable competition among medical practitioners: the printing press, which made it only "a small step from purchasing a remedy from a pharmacist to buying a book of remedies to cure oneself" (Eamon 138). Plants and books were linked more closely than before by the telescoping of the traditional medical hierarchy that was enabled by print. As a popular kitchen art and a highly rarefied academic discipline, sixteenth-century medical herbalism was bookish at both ends of the spectrum.

Words and Things in Humanist Herbalism

It was not only manuscripts of theology, politics, and poetry that were edited so assiduously for authenticity by sixteenth-century humanists. Because of its classical authorial forebears—Dioscorides, Pliny, and Theophrastus—herbalism was subject to the same international program of editorial intervention. Sachiko Kusukawa dates the genesis of a humanist herbalism from Niccolò Leoniceno's *On the Errors of Pliny and Others in Medicine* (1492–1509) attacking the Roman Pliny in favour of Greek authorities ("Leonhart Fuchs" 415–16); likewise, Allen Debus observes the application of a "literary humanism" to scientific texts in the later fifteenth century and to medical texts in the early sixteenth (132). This bookish botanical revolution was a necessary precedent to the turn to plants themselves. As one historian of the early modern naturalists put it:

> [B]efore the first experimentalists, Vesalius in anatomy and the herbalists in botany, had effected any serious change, there came the makers of commentaries who strove to purify the text[s] ... of Greece and Rome, and the makers of pandects who collected and collated the utterances of the ancients. (Raven 22)

Yet the return to books was not a turn away from nature, since such pandects, rather than embalming non-textual objects in textual form, led to a renewed and "more active engagement with the empirical world" (Peter Harrison 78).

The work of Pier Andrea Mattioli vividly embodies the humanist spirit of botanical scholarship: his life's work was a commentary on the text of Dioscorides, the chief classical authority consulted by Renaissance herbalists. Its popularity was astonishing: the *Commentarii* of 1544 was reissued in many editions and translations, including Italian, Latin, German, French, Spanish, Czech and possibly Hebrew. Mattioli claimed it sold over 30,000 copies by 1568 (Findlen, "Formation" 373–4). Like many authors of contemporary gardens of verse, Mattioli spoke of himself in his preface as tending "a vast and long-neglected garden, filled with broken, tangled, and unappreciated plants" (Findlen, "Formation" 375). The book, like the plants it described, grew in girth with each new edition. Its successive editions also, unexpectedly, show an emerging emphasis on original authorship and away from the respectful translation of and commentary on an *auctor*: the first edition was more or less a translation with very little commentary, while in later editions Mattioli's contributions expanded to cover not just plants in Dioscorides but all of the additional plants in his own acquaintance, until his own writing eventually "overwhelmed the text of Dioscorides" (Johnston, *Cleveland's* 10). His commentary on Dioscorides' short foreword, for instance, grew to a single paragraph of 14,000 words (Blunt and Raphael 133), and in some editions, the original text was excluded altogether so that the commentary stood alone (Johnston, *Cleveland's* 10). Such scholarship may have first led to a more rigorous approach to classical texts, but it also, clearly, could lead away from those forebears to the authoring of original texts based on the herbalist's own experience. Similarly, improvements in classical texts led, eventually, to a desire for improved vernacular texts, just as a better understanding of flora from classical locales led to a desire to understand home-grown vegetation. As Findlen writes, "[s]eeing nature and reading Mattioli's commentary became an ongoing, interactive process that epitomized in many ways the humanist ideal of a living text" ("Formation" 377)—an ideal that would do away with one of the basic boundaries between art and nature.

Yet the humanist ideal of a living text was not easily realized, and the interface between plant and text not unproblematic. The difficulty of making plant and text meet can be perceived at the site of their most basic interaction: the proper noun. Problems with nomenclature (or the lack thereof) were the thorniest in the transmission of natural-historical knowledge, so that much botanical work consisted simply of trying to assign correct names to plants. There was, after all, no standard naming system such as that which Linnaeus would bequeath to future taxonomists. The problem had confronted medical practitioners for centuries, but vigorous international efforts to address it only arose in the late medieval and early modern period (Siraisi 143). When they did, it rapidly became apparent that most of the world was a parody of an archive filled with apparently unlabelled botanical objects, while some regions, particularly the lands surrounding the Mediterranean, held collections of plants that had been labelled, relabelled, and mislabelled in a plethora of languages since antiquity. As the editor of the second edition of John Gerard's herbal wrote in his sympathetic note "To the Reader": "nothing more troubles such as newly enter into this study, than the diuersitie of names, which sometimes for the same plant are different in each Author; some of them not knowing that the plant they mention was formerly written of, name it as a new thing; others knowing it writ of, yet not approuing of the name" (Johnson). Similar problems were diagnosed fretfully by Francis Bacon in his *Novum Organon*: "[There] are either names of things which do not exist (for as there are things left unnamed through lack of observation, so likewise are there names which result from fantastic suppositions and to which nothing in reality corresponds), or [there] are names of things which exist, but get confused and ill-defined and hastily and irregularly derived from realities" (342). The consequence of this byzantine tangle of words and things resembled something cancerously organic: "[t]he number of plants grew, but the number of names grew even more quickly" (Ogilvie, "Many Books" 34).

Despite the absence of an overarching nomenclature, moreover, individual plant names were invested with a sometimes bewildering degree of significance. Some apparently saw names as a kind of key to unlock the code in which the Book of Nature was written, or even their own personal books of nature, the cabinets of curiosities compiled with such fervour in this period. The Italian botanical gardener and naturalist Ulisse Aldrovandi kept a collection of *naturalia* that included over 18,000 items arranged in books, cupboards (14 of these), armoires (66 of these), and pigeonholes (over 4,500). After detailing this remarkable number of locations, Aldrovandi concludes blandly by noting that the items are "marked with their names, so that they can be found" (qtd. in Swan 110). Only names stood between the collection and its loss—yet names themselves were in a state of chaos. Worse, the varieties of confusion were almost sufficiently numerous to require a nomenclature of their own, while the errors and misunderstandings perpetuated by the very glossaries and synonymies produced to sort out such matters often only added to the confusion (Tony Hunt xvi–xvii). For instance, the innocuous sea-urchin (*Echinus*) and "the ship-stopping remora" (*Echineis*) were conflated when their names were transliterated poorly; elsewhere the word *Ireos*, the genitive case of Iris, was not read as such but as a distinct species (Peter Harrison 72–3). In the first case, two species were reduced to one through the conflation of their names; in the second, one species was speciously multiplied through the mistaken multiplication of a name. Old names were trouble, but so were new ones, as distinguishing adjectives and other qualifiers were added to yield "at first binomials and, with the passing of time, increasingly lengthy and ungainly names" (Ogilvie, "Many Books" 33). In a medical text published as late as the *Botanicum Monspeliense* (Lyons, 1676) may be found names of up to nine words (Ewan 11). With every word added to a proper noun, the distinction between that category of language and ordinary discursive description grew less legible and less functional.

Trouble also arose when old names were matched with new plants based on false assumptions about the universality of the earth's ecosystem: classical sources described the flora of the Mediterranean and near East, but some medieval and early modern herbalists forcibly applied what they learned in those texts to plants in their own more northern and western habitats (Tony Hunt xvi–xvii). When writers did turn with an open mind to examine the vegetation in their own backyards, they confronted still more nomenclatural confusion, since such plants were identified by a plethora of local and dialectical aliases (Tony Hunt xvii). Meanwhile the utterly different ecosystems of the Americas and elsewhere led to an entirely different category of bewilderment. An instructive example is found in Gonzalo Fernández de Oviedo's *Historia general y natural*, translated by Richard Eden in 1555 as *The Hystoire of the Western Indies*. Oviedo's explanation of the initial reaction of the Spanish "highlight[s] the impossibility of naming most of the things their eyes were seeing. The most obvious feature of the newly discovered lands was the insurmountable illegibility of the landscape: 'Since, although it is visible, we ignore most of it, since we do not yet know either the names nor the properties of such trees'" (Castillo 491). Without names, the trees were hardly perceptible; the result is a text that often does not read like a descriptive natural history, as "the insufficiency of language to name things triggered a tumultuous verbal display organized around the sensorial immediacy of the natural phenomenon—its sheer presence—and the intense subjective affections it produced in the privileged beholder—his own experience" (Castillo 491). Oviedo's experience of the landscape triggered something like a lyric poetry in the midst of his natural history.

Humanism and herbalism may have been a peculiarly suitable match owing to the importance of nomenclature in the humanist approach to education. As Comenius wrote in the introduction to his *The gate of tongues vnlocked and opened, or else, A seminarie or seed-plot of all tongues and sciences* (1633), "If thou askest, whats to be learned? haue for answer, *To know the differences of things, and to be able to give its name to every thing.* Is there nothing more? Truly nothing at all. For he hath laid the foundation of all learning, that hath learned the surname of things" (B1r). In his address to the reader, Comenius links his botanical titular metaphor to the very same matter: "we thought good to intitle this our worke with the name of a Seminary or Seedplot, because here is an equall respect both of words and things" (A2v). That balanced ecosystem of words and things was not found extant in nature, however, but laboriously cultivated; and it was first cultivated in the sixteenth century by bookish sorts trained in the humanist methods of textual collation. The collection and recollection of books and the texts and names contained in them had to be undertaken before a renewed program of collecting and recollecting plants themselves could make much headway. Elizabeth Eisenstein's remark is pertinent: "Sixteenth-century investigators had to be concerned with ancient languages and inscriptions, with 'names and words,' whether their interests were 'literary' or not" (483). Yet the same investigators had to be concerned with things themselves in order to fulfill their work with words. When confronting a mysterious botanical name, "the only real option for the conscientious translator was to go into the field and match a known species with the description" (Peter Harrison 78). This option, driving the translator out of the study and into the garden, was a remarkable advantage for the humanist herbalists over their more literary colleagues: "where philologists usually had to rely on their knowledge of ancient languages, style, literature, and history to make their emendations, with botanical treatises they could make sense of corrupt passages by inspecting the living plants in question" (Reeds 27). A living link, in vegetable form, could thus be forged between ancient and modern texts.

Botany by the Book

Aside from the basic task of aligning *res* and *verba* so that each plant had a proper name, humanist botany was a remarkably text-driven and book-based activity. Herbalism was conveniently amenable to collaboration by correspondence involving a cross-pollination of botanical and textual materials. Unlike most animal or mineral materials, dried and flattened plant materials fit easily into a packet of letters, many of which were exchanged by botanical practitioners (Reeds 109; Morton 153). As Reeds puts it, "[t]he exchange and discussion of botanical books was as common in the letters of Basel doctors as the exchange of seeds" (109). Plants and paper goods thus became almost interchangeable currencies in this community. Such acts of correspondence were not merely sociable but critical to the production of botanical knowledge in the sixteenth century. Many of the illustrated items in Conrad Gesner's collection of some 1,500 samples were modelled on dried specimens sent by a European network of correspondents (Atran 298). Mattioli revised his commentary on Dioscorides based on both specimens and information received by correspondence (Eisenstein 110); among these specimens were those of Luca Ghini, who maintained voluminous correspondence with other prominent naturalists (Arber, *Herbals* 139–40). A little later, the herbalist Carolus Clusius kept up an international network of over 300 correspondents (Fat, "Clusius" 8). John Rich, Hugh Morgan and other apothecaries of the late-sixteenth century extended the scope of their physic gardens by exchanging plant material with continental experts (Raven 192). Botany by correspondence, then, was another way in which herbal and textual cultures were intertwined, one that helped make botany an early model of international collaborative science.

Gerard says of a plant he calls Sesamoides, "I haue had the seedes sent me from Padua in Italie" (*Herball*, 397), and notes that such a gift had the uniquely pleasing quality of being the sort that kept on giving: "The flowers I do expect this present yeere" (397). After a successful season, Gerard could collect the seeds and distribute them still further; similarly, the letters with which such seeds were sent could be copied and diffused among a wider circle of friends (Kusukawa, "Incunables" 118). Plant and text were susceptible to similar procedures of reproduction and distribution. The sharing of plants in this way can be seen as a way of turning private property into a kind of non-textual commonplace: like a quotation copied from one book into another, a plant could be given to another gardener without the giver losing his hold on the same. Gerard, moreover, shows that bits of text could be treated as a similar kind of gift when he recounts in his herbal that the German herbalist Joachim Camerarius "himselfe did giue vnto me at his being in London" a *description* of the Germain Flower-de-luce—not the plant itself, but the text describing it (50). On a larger scale, portions of herbals were copied from one to the next for the sake of economy or comparison.

The humanist herbalist's habit of commonplacing in this way was taken up in a more popular medium through the collection of medical prescriptions, most of the ingredients of which were botanical; these were, like commonplace quotations, texts that "were freely copied and circulated by patients and practitioners alike" (Pelling, *Medical Conflicts* 109). Pelling notes that "[t]he end product of this process of circulation could often, of course, be receipt or remedy books" (109)—books whose resemblance to and tendency to overlap with commonplace miscellanies has often been noted.[4] Just as commonplace scraps of verse were often treated

---

4 Marotti refers to manuscript poetical anthologies including "catch-all miscellanies containing such other material as prose letters, medical recipes, household accounts, school

as anonymous texts, "such bills failed to preserve details like the physician's name, the date, or the name of the patient" (110). And just as the commonplace tradition that first developed through the circulation of manuscripts later adapted itself to print culture, the copying of medical bills "naturally extended to extracts from printed sources or collections, the copied versions then taking on an often anonymous life of their own" (110). The collaborative, commonplacing culture of exchange in herbal and verbal materials was clearly not limited to the humanist circles that encouraged the keeping of commonplace books.

The type of text most memorably associated with sixteenth-century botanical studies, however, is the printed herbal: an encyclopedia, often richly illustrated, and consisting primarily of prose describing the appearance, habitat, known names, and purported "virtues" or curative powers of plants. The herbals remain some of the more familiar early modern books, and they made their mark on contemporary audiences, too. Debus contends that "the herbals were among the most popular books printed in the sixteenth century" (48). The book known as Banckes's herbal, the first printed in English, reappeared at least 15 times between 1525 and 1560 (Bennett 98). The wherewithal of some patrons could be an influential factor in the production of expensively illustrated folio herbals (Frank Anderson 4), but according to a historian of the book trade, the success of sixteenth-century herbals can be attributed to a wider market (Bennett 109). While many herbals were created by academically-trained medical men for their own ilk, some first aimed at professionals were later adapted for a more popular audience, either through translation or the publication of smaller, more affordable, and often pirated editions (Arber, *Herbals* 6–7; Henrey 36). The printer's calculus was simple: everyone gets sick, so everyone is in the market for a herbal. The same logic explains the analogy of the devotional herbal in a number of contemporary titles; the broadly inclusive demographic of the spiritually ailing were served by Christian pharmacopeias marketed under names like *The Divine Herball* (1616), *A garden of spirituall flowers* (1609), or *A sweet posie for Gods saints, containing many choyce, and sweet flowers gathered out of Gods owne garden, the holy Scriptures* (1642). To a considerable portion of the population, a herbal became "invaluable" and "essential" (Henrey 5).

Before it had become so, though, the printed herbal needed to be made familiar to its potential buyers. People commonly kept household books in manuscript form for storing herbal recipes or inventories (Hoeniger *Medicine* 17–18; Havens 9), but their connection with printed artifacts in a bookseller's stall might not have been immediately obvious. Even fifteenth-century printers had trouble characterizing their herbal works; Arber notes with respect to incunabula that "if a herbalist of that date had wanted a specific title for his book, it would have been by no means easy for him to find an appropriate term, since botanical language was in its infancy" (*Herbals* 17–18). Not just botanical nomenclature, then, but

exercises, and journal writing" ("Malleable" 162). Earle Havens notes that "medicinal and culinary recipes" are frequently intermingled among the entries in early modern commonplace books (9).

biblio-botanical nomenclature was inchoate before the sixteenth century, and the bibliographic consequences paralleled those in botany, as titles multiplied well beyond the number of objects needing names. The Latin *Herbarius*, for instance, also came to be known as *Herbarius in Latino, Aggregator de simplicibus, Herbarius Moguntinus*, and *Herbarius Patavinus*, while the German *Herbarius* may be found under aliases including *Herbarius zu Teutsch, Gart der Gesundheit*, and *Ortus sanitatis* (Arber, *Herbals* 20, 22). One all-purpose solution adopted by printers was to characterize herbal works, as some of the examples above suggest, by botanical metaphors that construed the books as gardens (*Herbals* 17–18). The analogy transplanted a conventional metaphor for poetry into the field of herbalism. The ensuing popularity of printed herbals in the sixteenth century may, in turn, have led to a more vigorous exploitation of that same metaphor by more literary publishers who used the trope to suggest the curative or, in the language of the time, the virtuous properties of their gardens of verse.

It was not just titular imagery but actual images that made herbals both popular and ground-breaking books. Indeed, the first sign of the sixteenth-century botanical renaissance is often identified in the 1530 herbal entitled *Herbarum Vivae Eicones* (Brunfels), a book that made its mark with its eponymous "living images of plants" by Hans Weiditz. In this book, the humanist ideal of a living text expanded to encompass visual representations that made the accompanying text exponentially more vivid and useful. Of course, many manuscript herbals had images, and early printed herbals often imitated them in woodcut form, in which the transition to the less sensitive medium of the woodcut could entail a step backward in naturalistic representation. Many botanical images in early printed herbals were "symmetrical, schematized, structurally ambiguous decorations" that were "identifiable only if one already knew the plants they were supposed to represent, sometimes not even then" (Reeds 31, 28). This is in part because so many were, like scraps of verse appearing over and over in commonplace books, simply plucked from their original context and placed in another textual habitat, rather than being based on actual specimens.

In the "living images" mentioned above, on the other hand, the artist worked not from models but from living or recently gathered plants. The result was not perfunctorily realistic but resulted in a refinement of the woodcut technique so far as to reproduce the idiosyncrasies of individual specimens, including broken stems and wilted or folded leaves (Hoeniger and Hoeniger, *Development* 26; Blunt 47). In this way the images can closely resemble the plant material that began to be pressed and kept in albums called herbaria. Rather than create a generic, composite, or idealized image, Hans Weiditz represented unique specimens, not the species. Ironically, print reproduced their idiosyncrasies—not just in the original editions but in later books that copied the designs—to the point where they appeared representative of something essential about the plant. These images could thus be said to be *vivae* in another sense: the images were subject to the same rapid and high-fidelity reproduction that was one of the most valued qualities of plants themselves. Even Weiditz's most moribund plants were given unprecedented new life by the reproductive powers of print. The fidelity of print reproduction was, of course, as Adrian Johns has recently made clear, still subject to considerable mutagenesis; indeed, one of the more ironic ways in which print advanced botanical studies was by drawing attention to the corruption that could occur when images were copied, simply by accelerating the rate at which they were copied and, hence, the rate of deterioration (Eisenstein 108). For instance, Fuchs's herbal of 1542 was adapted and copied in at least a dozen editions before 1550, including an edition of 1545 that shrank the images so much as to make them barely useful, while those from 1550 and 1551 "compromised botanical integrity, as flowering parts disappeared, leaves lost their veins, and other details were sacrificed" (Hunter 10). The printing press, far from being exclusively an agent of fixity, could indeed be an agent of change.

After Brunfels and Weiditz, the herbal attributed to Leonhart Fuchs is recognized as having taken the next significant step in botanical illustration. The images in Fuchs' *De Historia Stirpium* differ from those in Brunfels' *Eicones* in being composed of fine outlines, without Weiditz's three-dimensional shading; moreover, the images do not portray the accidents of individual specimens but endeavour to represent the plant in its most typical or perfect form (Kusukawa, "Leonhart Fuchs" 403). Imperfections assiduously included in the Brunfels-Weiditz production were just as assiduously eliminated from the images in Fuchs. The results, however, could be idiosyncratic in their own way, since in some images the bud, flower, and fruit that appear sequentially in nature are portrayed as if they arose simultaneously (Blunt and Raphael 142). Later in the century, the same visual convention was taken up in herbals by Clusius, Dodoens, and l'Obel, prominent herbalists who shared a collection of woodcuts which present both fruit and flower (Blunt 68). This choice may have been merely economical (Blunt 54) but also happens to reflect an aesthetic ideal frequently alluded to in contemporary verse: the desirable impossibility of combining flower and fruit, pleasure and profit, at once.[5]

Objective naturalism was thus not the only ideal in herbal images; instead, aesthetically motivated choices were often made by botanical artists. Arber notes that "the herbal illustrator who drew upon the block" was often "so much obsessed by its rectangularity that he accommodated his drawing to its form in a way that was unnecessary and far from realistic" (*Herbals* 215). In his 1588 herbal, Tabernaemontanus split a single stem in two in order to fill the woodblock, or elsewhere "stiffened the plant out, and gave all its branches the same height," in either case "rendering the figure very suitable for use as ornament" while blithely sacrificing what we would see as the scientific virtue of its resemblance to the specimen or species represented (Hatton 3). In the visual interrelation of plants and print, the artistic medium of the woodblock could thus become a large part of the botanical message. That the ornamental quality of a botanical image should

---

[5] The prefatory verse to Thomas Nicolls' translation of Thucydides, for instance, celebrates an achievement in which "The frute wyth the flower herein is comprised / Wyth eloquence" (Ai^{v}).

matter considerably to the maker of a herbal or its audience becomes evident in Jerome Bock's herbal of 1546, illustrated by David Kandel in part with wood-cuts of pastoral scenes one might expect to find instead in a book of poetry (Figure 2.1); there is no botanical reason for the illustration of the mulberry tree, for instance, to include a representation of Pyramus and Thisbe (Frank Anderson 134). These images, with aesthetic qualities that defy exclusively technical purposes, are thus among the clearest demonstrations of the early modern imaginative interpenetration of what we would distinguish as the scientific and the literary.[6]

In another way, too, the composition of the herbals resembles that of many poetic texts of the period. Although often known by the name of a single author, the herbals were collaboratively produced by many hands (Findlen, "Formation" 373). The anthological nature of what is called Fuchs's herbal can be seen even in the list of its artistic contributors: Albrecht Meyer drew the plants that Heinrich Füllmaurer copied onto the block of wood before Veit Rudolf Speckle cut the design. Their contributions were thought worthy of sufficient respect for their portraits to be included; the trio appears on a single page, as if better to express their interdependency (Figure 2.2). Despite the long-standing tradition of collaborative artistry, this may be the earliest visual record of the teamwork that went into the herbal illustration (Hunter 9), which suggests a heightened awareness in the sixteenth century of the importance of the anthological creative model in the herbals. Not only was collaboration among one or more herbalists, a number of artists, and a printer a practical necessity; even after the point of sale, the buyers of the books were encouraged to participate in the production of the herbals by colouring in the images (Blunt and Raphael 115) or by responding to the author's requests for corrections, a notable feature of many herbal prefaces.

That collaborative model was echoed in the contemporary culture of English medical books, which a historian of medicine has characterized in a similar fashion: "A large number of minor figures, working in close collaboration with printers and publishers, produced translations and paraphrases of continental works for the English market, often adding their own material or urging attention to the native tradition" (Pelling, *Common Lot* 35). In other words, English medicinal herbalism is most recognizable for its anthological textual productions. The same historian goes on to note that "parallels could be drawn with respect to the response of English poets and dramatists to continental models, a process also involving minor figures and complex bibliographical problems .... The emergence of a literature in the vernacular was a major feature of both developments" (35). There was thus a marked resemblance between the period's botanical and literary developments.

---

[6] "No clear line of demarcation as yet separated cognitive and technical works from those we today designate as 'creative' and aesthetic: all found a place in the larger domain of literature" (De Bruyn 348). Thomas Moisan refers to this state as a "pre-Royal Society ... elision of the empirical and the imaginative, wherein curiosity over natural phenomena and attention to taxonomy are inflected by an apprehension of the supernatural and mediated by a vocabulary and affirmation of the uncanny and mythic" (310).

Fig. 2.1 One of David Kandel's illustrations in Hieronymus Bock's *Kreuterbuch*

Fig. 2.2 Portraits of the illustrators of Leonhart Fuchs's *De Historia Stirpium*

Some historians have gone so far as to say that what we would now think of as a literary mindset is required to understand early modern natural history: "If we wish to study [the naturalists] Gesner, or Aldrovandi, or Camerarius, and understand their works in the spirit in which they were written, then we must be prepared to think adagially, allegorically, and analogically" (Cook, "Emblematic" 36)—and also, it would seem, anthologically.

In this way the herbals bore a perceptible relation to the commonplace tradition, in which text is not imagined as a substance under sole ownership. Brunfels, for instance, though often treated today unproblematically as the author of the herbal known by his name, "was in fact little more than another of those industrious gleaners of miscellaneous material" (Blunt and Raphael 120); like so many other herbalists, he compiled a botanical version of a commonplace book. The same could be true of herbal images; the woodblocks of Clusius, L'Obel, and Dodoens were held in common by their printer Christophe Plantin, and the blocks became popular with others, as "is shown by the frequency with which they were copied" (Arber, *Herbals* 229, 232)—like commonplace pieces of verse appearing again and again in printed anthologies and manuscript miscellanies. Because the medium of the woodblock was so expensive and enduring, images were used repeatedly in different books; those from Fuchs' *De historia stirpium* were still usable two centuries later, and the images in Mattioli's *Commentarii* appeared in a French herbal of 1766 (Arber, *Herbals* 223). Similarly, the woodcut images of maize found in almost 50 herbals published in the sixteenth and seventeenth centuries derive from originals in just seven books (Stearn, "Use" 5). An anthologized poem or commonplace possesses a similar transferability and durability.

Botanical Gardens in Book Form: The Herbaria

While it may be argued that few advances in sixteenth-century botany were made outside of the collection, arrangement, description, and pictorial representation of plants in books (see Hoeniger, "Gesner" 82), an apparent exception to this statement may be found in the botanical garden. This new kind of garden, which expanded traditional physic gardens in an attempt to contain a specimen of every known plant, useful or otherwise, was first established in the 1540s in Pisa and Padua. These were followed rapidly by others at Florence and Bologna and finally by privately owned and often princely imitations outside of the universities and Italy. While these open-air archives may appear to confute assertions of the bookish nature of sixteenth-century botany, they in fact yielded a harvest not only of plants but of still more books. That is, botanical gardens engendered textual counterparts that acted as bookish mirrors to the gardens themselves, both in the relatively new form of the garden catalogue and the entirely new form now known as the herbarium.

The need to catalogue the extensive and proliferating contents of botanical gardens meant that a textual order was applied to the garden itself, most immediately and proximately in the form of beds that were often ordered and "distinguished by letters of the alphabet and the plants by numbers" (Tomasi 82). Writing on the land in this way visibly narrowed the gap between plant culture and textual culture. These letters and numbers were then made to correspond with those kept in a catalogue that recapitulated the garden in textual form. Like the more remarkably innovative herbarium, the innovation of the collection catalogue as a genre and a technology may be traced to this period (Findlen, *Possessing* 36; Swann 9).

Such catalogues were initially closely attached to their gardens as supplements to them, but before long they became texts of interest to a larger reading public (Findlen, *Possessing* 37). By the end of the century, a plant collector like John Gerard would find it worthwhile to arrange for the publication of a catalogue of his own private botanical garden, and to expand it and reprint it in a second edition only three years later (see Chapter 4). The intellectual aim behind the correlation of material and textual collections in botanical gardens has been described as "the recovery of knowledge, and of power over nature. By bringing all the plants ... into one place one could name them, and by naming them men could communicate the nutritive and medicinal properties of the plants to each other" (Prest 54). In other words, before knowledge could be power, language itself had to become more powerful; in this interpretation, the physical collection of plants was simply the premise for a secondary collection of names that was the actual if less spectacular point of the entire enterprise.

Far more innovative than the catalogue, however, was a type of book first brought into being shortly before the middle of the sixteenth century and, like the catalogue, first designed to act as a bookish counterpart to botanical gardens: this was the album of dried plants then known, rather poetically, as a *hortus siccus* or *hortus hyemalis*—a dried garden or a winter garden—and now better known as a herbarium (Raven 70). Both terms from the period demonstrate an explicitly and materially botanical, as opposed to a purely textual or literary, use for the metaphor of the book as garden. The herbarium was, moreover, an original adaptation of the form of the book as a tool for use in the study of plants. It consisted of a bound book with plants or plant parts first pressed between heavy paper, to flatten them, and then sewn or glued into place on its pages (Figure 2.3). As a so-called winter garden, it was useful to students otherwise unable to study simples in seasons when plants were dormant or dead; a deficiency in the archive of actual objects was corrected in a medium that had the power to represent plants at only one remove from their living being. As a dried garden, the herbarium offered a form of botanical mummification or taxidermy at a time when a comparable method of preserving animal flesh was not readily available, even though methods that preserved bodies after death were widely sought and, when they worked, "seemed almost miraculous" (Cook, "Time's Bodies" 226–9). The herbarium was thus a physical hybrid of plants and books, remarkable not only in retrospect but in its own day.

A sixteenth-century herbarium is portrayed in a painting of 1562 by Francois Clouet held in the Louvre; the portrait suggests the esteem in which the herbarium was held by its presumed maker and owner, the apothecary Pierre Quthe. Quthe's decision to be painted with his herbarium open at his side is a quirky modification of the conventional approach taken at the time to many portraits of herbalists and the iconic apparatus used to represent their occupation. Often, such portraits show a man with a book or a plant, or sometimes one in each hand.[7] Quthe's herbarium

[7] For examples, see, for instance, Arber's *Herbals* (58, 65, 81, 99, 131, 137).

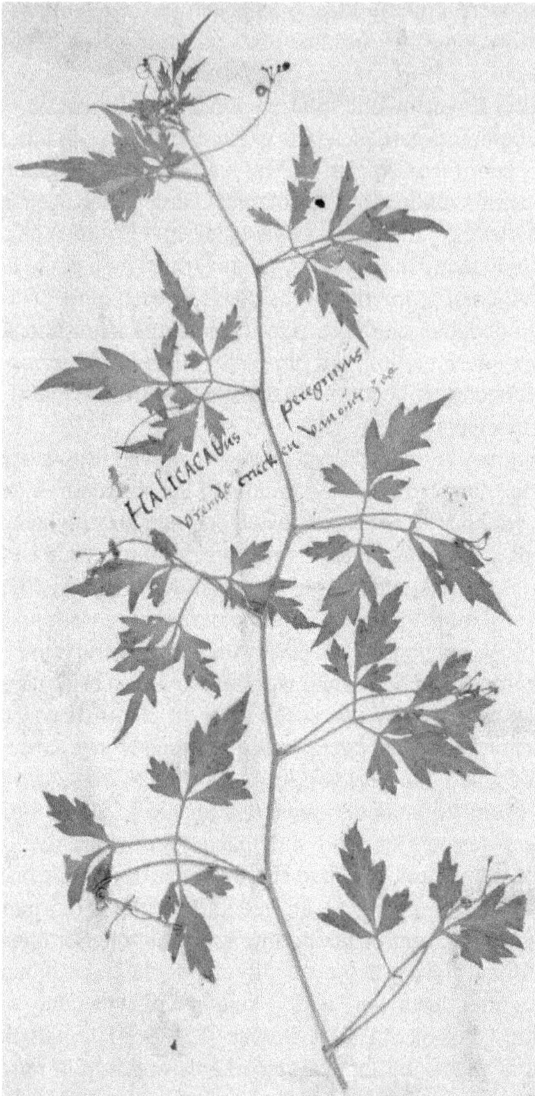

Fig. 2.3 A page from the herbarium of the apothecary Petrus Cadé. Reproduced by permission of the Utrecht Herbarium and the Petrus Cadé Herbarium website (http://www.bio.uu.nl/~herba/Cade/). Photograph by Renske Ek

is distinctive, and renders its bearer so, in its ability to combine the two disparate materials of plant and book as one; it suggests an achieved equilibrium between experience and textual authority in his learning as a herbalist. Moreover, because of its bibliographic format, the apothecary's hands are left noticeably free, and the relaxation that this lends to his posture suggests the practical advantages that such a technology would lend Quthe in the pursuit of his trade: whenever he wanted a particular plant, it would be at his fingertips, compacted and indexed in a way that let plants partake of the material advantages and capacities of the book.

The herbarium turned out to be, despite its humble appearance, an important advance in the early modern emergence of science; one historian of botany states that the herbarium "has been the essential basic tool of descriptive and systematic botany ever since" (Morton 123).[8] It has, nonetheless, not entered the popular imagination with the same iconicity as has, say, the slightly later innovation of the telescope in the early seventeenth century. Perhaps the conventional bookish form of the herbarium disguises its revolutionary nature as a fusion of plant material with the basic form of the codex. In the first half of the twentieth century, Agnes Arber referred to "the obscurity that still surrounds [the] early phases in the history of collections of dried plants" (*Herbals* 139); not much has changed since she wrote, perhaps not least because the subject can seem as dry as its materials. Yet the herbarium embodies in the most explicit form the way in which the early modern study of plants depended on books in unique and, in this case, novel ways. The very names used to refer to this new object (dried garden and winter garden) appeal to a metaphorical identity of the otherwise disparate material forms of book and garden. Even the early modern Latin term for a herbarium was the product of a conceptual crossing between plants and books: in medieval Latin, *herbarius* meant a written herbal, but in the early modern period it came to signify, instead, these collections of pressed plants in book form (Morton 153). But the herbarium did not only unite plants and paper metaphorically or linguistically; it was also a node at which the period understanding and use of plants and texts intersected materially.

Plants fused to the material of a book were instrumental in ways that living plants, whether in their own habitats or transplanted into a garden, were not. The practical uses of herbaria were many: they could be taken into the field, for example, and their contents compared with living specimens. They, or leaves extracted from them, could be incorporated into correspondence. They provided identifiable specimens out of season (hence *hortus hyemalis*, with its paradisal connotations of a *ver perpetuum*). Rare specimens could be preserved for many years and thus reach a wider "readership" than any short-lived specimen, which might not last a season in a garden. A book could be set among other books, and each herbarium could conceivably be shelved among others to create a comprehensive archive that would outdo the long-standing idea of the Book of Nature by transforming it into an entire library. The capacities realized in the herbarium through the material unity of plant and book may account for at least some of the connotations of the metaphors in sixteenth-century books that treated texts as flowers that could be easily carried about, dispatched across the country, or preserved long after the bloom had worn off, whether in a folded piece of correspondence or the

[8] Lorraine Daston's "Type Specimens and Scientific Memory" lucidly explains the role played by herbaria in the development of botanical knowledge.

closed pages of a book. But herbaria had yet another and much more remarkable function: they provided a crude means of suspending and then restoring at least the appearance of life. Mattioli referred to his own technique, when he desired to draw an exotic plant that had been preserved and transported in herbarium-form, of soaking the plant in water in order to restore something of its three-dimensional shape before drawing it (Blunt 60). This aspect of the herbarium is especially suggestive when considered in light of the literary metaphor of the garden of verse: like flowers in the herbarium, poems rendered two-dimensional by being placed on the printed page could be resurrected to a metaphorically three-dimensional form when projected into space by being read aloud; as one writer put it in a prefatory poem to *A Divine Herball*, "The Herbes which these dead leaues now bring, / Thy liuing voice did Sweetly sing" (A4r). The living and embodied voice of the reader could lift the poem off the page in the same way that plants embedded in the pages of a herbarium could be lifted up and infused with at least an appearance of bodily life.

Herbaria are thus notable for their unusual ability to narrow the gap between a representation and the thing represented, and they could therefore be said to problematize the broader gap between those overarching categories, art and nature, the relationship of which was of such overwhelming intellectual interest in the period (Daston and Park 260). The artful flattening of the plant or flower altered its natural appearance to the point where it could be considered a mere image of itself; at the same time, such an image was not nearly as far removed from the original as was a black-on-white woodcut print in a herbal—which was at four removes from the original (since the image for a woodcut was first drawn on paper, then traced onto the woodblock, then carved into the block, then printed on the page). The sheer immediacy of the sights enclosed in the herbarium suggests that they are still a part of nature, and that by opening the book one is entering a planted garden. Yet by being enclosed in a book, by fusing with the material of the page and thus losing a dimension, the plant material also becomes a representation of itself.

The herbarium was not the only way of playing with degrees of mediation between a botanical thing and its representation. An early-seventeenth-century florilegium, kept in the library at Blickling Hall in Norfolk and compiled by the naturalist Fabio Colonna, contains both paintings of flowers and "the impression of plants produced by pressing inked specimens" (Blacker 11), thus layering two levels of representation. The courtly botanist Paolo Boccone's herbaria, later in the same century, were also often produced from applying ink to plants and then applying the plants to the page. More remarkable are the pseudo-herbarium *trompe l'oeil* paintings made in the 1590s by Joris Hoefnagel at the court of Emperor Rudolf II in response to the designs of calligrapher Georg Bocskay's various scripts from the 1560s (Kaufmann and Kaufmann; Kaufmann, *Mastery* 14–15). Hoefnagel treated the image of a plant as visually congruent with the image of finely-painted words, as the unfurling tendrils of script are mimicked by the convoluted petals (Figures 2.4 and 2.5). Here, the materiality of the text, as script, is brought out and paralleled by an allusion to botanical material. But Hoefnagel goes further: to preserve the original calligraphic line in the middle of the page, he makes it appear as if the flower's stem punctures the page and passes through it, like a pin. Hoefnagel shows his eye for detail when he completes the effect on the reverse of the page by painting the appropriate glimpse of stem that is supposedly piercing it. Herbaria, like these images that mimic them, could similarly toy with the relationship between art and nature in a way that would contribute to their broad intellectual and aesthetic appeal at a time when such hybridity was a valued quality and a salient feature of objects considered worthy of collection. This, in turn, may be yet another reason that it is so often specifically in textual *collections*, far more than in other sorts of books, that metaphors of books as plants tend to occur.

Just as the herbarium exploited material qualities of the book in the service of the study of plants, it can be said to have exploited material qualities of plants in the service of a bookish culture. The specificity of this book-plant connection becomes clearer if one turns, for comparison, from herbals to bestiaries and seeks an equivalent to the herbarium in relation to the animal kingdom: that is, a medium in which the organism could be artfully represented with the same minimum of mediation. The cabinet of curiosities, with its stuffed creatures affixed to the architecture, comes to mind, as do the reptile- and amphibian-embedded ceramics of Bernard Palissy, who undertook his most famous work around the same time as the herbarium was having its heyday. Palissy is remembered for his ceramic platters and bowls, to the surfaces of which he would fuse casts made from fresh specimens of lizards, frogs, snakes, and molluscs (Figure 2.6). His glazing would attempt to reproduce the often jewel-like but naturalistic coloration of their anatomies (Lecoq 75). Palissy tried to fuse nature and art in the form of animals and pottery, just as the herbalists did with plants—but the artful form to which the latter turned was the book. The relative freakishness of Palissy's creations highlights the distinctly different successes of the herbarium in fusing organic and artistic material. Nature and art meet in both forms, at the same point in time and in the same intellectual climate, in the most literal way imaginable. But the purpose and effect of Palissy's work is aesthetic: the result is affective, yielding wonder or perhaps disgust. The purpose and effect of the herbarium is very different. Where the contents of Palissy's works appear chaotic and almost overwhelm the forms that try to contain them, the snakes all but slithering away, the herbarium appears orderly and contained by its form. The codex format entails a distinct beginning and end, while Palissy's round dishes offer no linear direction to guide a viewer from start to finish. The herbarium has a clear purpose, where Palissy's tumultuous surfaces vitiate the typical purposes of similar dishes. The conspicuous uselessness of his particular combination of *naturalia* and *artificialia* highlights the oddity of the herbarium in achieving a working unity between the two.

While the herbarium was an instrument designed for archival and educational purposes, it could also be put on display, as the portrait of Pierre Quthe demonstrates. The herbarium of Felix Platter, a Basel physician who began collecting as early as 1554, was open for viewing by distinguished visitors; Michel de Montaigne

Fig. 2.4 A page from Joris Hoefnagel and Georg Bocskay, Maltese Cross, Mussel, and Ladybird, Fols. 1–129 written 1561–62, illumination added about 1591–96

recorded the pleasure he felt when he saw it while travelling through Basel in 1580 (Reeds 108). Paolo Boccone, mentioned above, would create herbaria as gifts for his patrons (Boutroue). Not only were herbaria collections in themselves, then: they too were valuable as collectible objects. Via the herbarium, individuals could participate in the artful displays of curiosities that were so important a part of elite collecting culture of the day.

Botanists of the eighteenth century determined that the advantages of the book form were outweighed by its disadvantages and abandoned the bound format of the herbarium in favour of a system that would accommodate an unlimited number of species, as well as their ordering and reordering. By breaking the spine that held the arrangement of plants in a finite sequence between two covers, they were able to expand a single herbarium's empirical potential infinitely. In place of the book as the governing structure of the herbarium appeared the technology

Fig. 2.5 The reverse of the same page from Joris Hoefnagel and Georg Bocskay, Trompe l'Oeil Stem of a Maltese Cross, Fols. 1–129 written 1561–62, illumination added about 1591–96

of the filing cabinet, a more modern knowledge-management technology that was only recently surpassed or at least supplemented by the networked electronic database. Today's herbaria, in their material form, bear almost no resemblance to the bibliographic herbaria of the sixteenth century: they consist of whole climate-controlled warehouses of thousands of plants and plant parts of historical and contemporary significance. The early-modern bibliomorphic herbarium does not express some natural or necessary concomitance between the material forms of book and plant; rather, the changing shape of the herbarium over time shows that the convergence of book and plant in the sixteenth century manifests a local concentration of relations between these two objects.

While many leading herbalists seem to have at least been aware of herbaria, the technique of making one was not published until 1603. In that year, Adrian van Spieghel printed the first detailed instructions, which demonstrate what

Fig. 2.6 Bernard Palissy, Oval Basin, about 1550

a labour-intensive and expensive process it would be—combining the daily labour of planting and harvesting a garden with the material costs and processes associated with printing a book (Morton 153–4). Spieghel notes that plants must be compressed between layers of high-quality paper, with weight being added slowly, and the plants examined and turned over each day until dry, at which point they are to be laid upon separate pages and painted with a special glue, then laid upon more paper, shrouded with linen, and rubbed until they stick to the page. The final step involves still more paper when the pages with the plants impressed upon them are placed between other pages, or in a book, and once again compressed until the adhesive dries (Arber, *Herbals* 142). Because paper was not an inexpensive commodity—accounting for up to three quarters of a book's cost (Watt 1)—the process of constructing a herbarium would have been a relatively expensive one, and the result a precious product; its luxurious nature would contribute, again, to its evaluation by collectors who admired such lavishness.

It is the academic Luca Ghini who is usually attributed with the invention of the herbarium (Raven 77); the oldest surviving herbarium belonged to Gherardo Cibo, one of Ghini's students (Arber, *Herbals* 139). Because of Ghini's work as a teacher, his technique was rapidly disseminated across the continent when his students went on to teach others the technique (Reeds 35); thus, although Ghini is unusual among the botanical community of the period in that he published no books, he still promulgated a novel use for the book form that provided an innovative link with plants. Among the earliest users of the technology were two

English men, John Falconer and William Turner (Raven 77). The first published evidence of a herbarium is in reference to the otherwise elusive Falconer, a figure mentioned frequently by the Portuguese physician Amatus Lusitanus, who in 1553 described Falconer as "a man fit to be compared with the most learned herbarists, a man who had travelled many lands for the study of plants and carried with him many specimens ingeniously arranged and glued in a book" (translated in Chapman and Tweddle 19). Turner himself refers to "my frende Falconer" and to "Maister Falkonner's boke" in his 1562 herbal, and to Falconer as one among a group of prominent and learned English herbalists (Raven 69, 77–8). Falconer, like Turner, travelled in Italy and likely also learned to make his herbarium from Ghini himself (Arber, *Herbals* 140). Turner studied with Ghini in the early 1540s, when Ghini's herbarium must have been well under way. Around then, Turner must have constructed his own herbarium, since he refers in the second volume of his herbal, published in 1562, to leaves of the "Fistick Nut whereof I have certayn at thys day to shewe well kept in a booke at the lest these seventeen years" (509). The value at the time of any such book in advancing botanical study is made clear by the following comment in the same volume, made in relation to Falconer's herbarium: Turner writes of a plant named Sea Trifoly, "I never sawe it in Englande savinge in Maister Falkonner's boke and that had he brought out of Italy" (392).

After the adoption and dissemination of the herbarium technique in the middle of the sixteenth century by serious students of botany, simpler approaches to preserving flattened plant parts also came to be recommended, for instance, by Hugh Plat in his *Delightes for ladies*, in which he suggests a method that he feels is "very requisite for a good simplifier"; his method joins the genre of the herbarium to that of the printed herbal: "he may drie the leafe of any hearbe ... and lay it being drie in his Herball with the simple which it representeth, wherby hee may easily learne to knowe the names of all simples which he desireth" (Item 63). Herborizing activities inspired by the form of the herbarium could be not only a delight for ladies but a respectable part of a gentleman's education around the same time. Writing in retrospect of his own education in the late sixteenth century, Edward, Lord Herbert of Cherbury, praised herbal studies that united the materials of the printed herbal with plants in the field in a way that seems inspired by the herbarium:

> I conceive it is a fine study and worthy a gentleman to be a good botanic, that so he may know the nature of all herbs and plants, being our fellow creatures and made for the use of man; for which purpose it will be fit for him to cull out of some good herbal all the icons together, with the descriptions of them and to lay by themselves all such as grow in England, and afterwards to select again such as usually grow by the highway-side, in meadows, by rivers, or in marshes, or in cornfields, or in dry and mountainous places, or on rocks, walls, or in shady places, such as grow by the seaside; for this being done, and the said icons being ordinarily carried by themselves or by their servants one may presently find out every herb he meets withal ... (Herbert 56)

Herbert's creative pedagogical exercise illustrates an evolution in plant culture and print culture over the course of the sixteenth century: where earlier plant collectors had to "cull out" their specimens from the land and put them into books, the process could reverse itself by the end of the sixteenth century, with the amateur extracting the images of plants out of books and returning them to their place in the land. The technology of the book had been successfully marshaled in the aid of archiving botanical knowledge in the sixteenth century; yet Herbert's project also demonstrates the limitations of a book as a way of organizing knowledge. He instructs his student to cut out and rearrange the images, obliterating the typical sixteenth-century herbal's alphabetical order for one that privileged habitat. His adaptation of the herbal is another sign that the coincidence of books and plants that was so marked in the sixteenth century was neither natural nor necessary, any more than it was permanent.

The genre of the herbarium is a most unusual kind of book, one that fuses plant and book in the most literal and immediate way. The very material nature of the book—including the thinness and flatness of its pages, their ability to link along a spine, the pressure they exert upon one another when enclosed between two covers, and their ability, when shut, to exclude the entry of light (and thus to slow organic decay)—was combined with the material nature of the plant, especially its susceptibility to a similarly page-thin flatness. What this meant was that a record-breaking minimum of mediation was at work in this medium, one that all but transformed the representational powers of the book into presentational powers. Howard Marchitello similarly suggests that the late seventeenth-century herbarium of John Evelyn presents itself as "fully in the world of a new georgic idiom, one that offers not signs or representations, but rather the very things themselves, an idiom whose referent is the natural world and not what the natural world is thought to represent" (166). Maria Zytaruk writes with respect to the same album "[t]his kind of collection demonstrated that 'reading the book of nature' did not have to remain a metaphor" (189). Indeed, by means of the herbarium, the existing affinity between the material form of the book and that of the plant was exploited and made stronger. This same affinity was exploited elsewhere in the same period, not materially but metaphorically, and not in a botanical but in a literary context. The herbarium thus embodies an important refraction of the trope of the early modern garden of verse, and perhaps best exemplifies the more broadly and uniquely bookish nature of sixteenth-century botanical culture.

Chapter 3
Botanical Reformation in William Turner's Books of Nature

Although significant advances in plant studies were made on the continent in the sixteenth century, William Turner appears to have been uniquely vigorous and productive among the English in his participation in plant culture. The advances he made depended not only, however, on his forays into English forests and fields in search of uncatalogued native species, as is often picturesquely highlighted. Instead, Turner's botanizing depended just as much on a return to and a new recruitment of the representational forms, qualities, and capacities of the book. Moreover, his work as a botanist depended on and was integrated with his work as a church reformer at a time when such reforms were deeply entrenched in textual controversies. This chapter explores the integration of Turner's understanding of plants and his approach to textuality by considering his work as a version of the contemporary botanical renaissance that, in light of his theological commitments, might better be described as a botanical reformation.

Turner's comprehensive deployment of old and new bibliographic technologies in the service of botany encompassed both print and manuscript as well as private and published works. Among his many other unrelated publications, Turner composed or compiled a remarkable number of works in the field of herbalism, with a total of five printed volumes surviving as well as a manuscript herbarium that has been lost. Turner also apparently wrote *De Naturis Herbarum*, which may be the Latin herbal to which he refers in the preface to his *Names of Herbes*; one might also note his role in a non-botanical anthology that partakes of the metaphor of the floral text under the name of *Sententiarum Flores* (Raven 71). He published his first botanical book in 1538, when only four other books on plants appear to have been printed in England (William Thomas Stearn 3), books that have been characterized in a comprehensive bibliography of such literature as "displaying botanical knowledge at a very low ebb" (Henrey 21). Turner's first publication on plants was therefore a decisive development in England, since the same bibliographer goes so far as to call it the opening of "a new epoch in botanical literature in England" (21).

As if reflecting the developing potential of print, each of Turner's botanical books is almost exponentially larger than the last, with his 1538 publication of a mere eleven pages succeeded by the 62-page book of 1548, both of which were thoroughly outdone by the three-volume folio herbal completed in the year of his death. In this way, the trajectory of his career can be seen to parallel the Virgilian literary model of authorship, with an apprenticeship in modest pastoral preceding the epic masterpiece (of a scope to which Turner's massive herbal is certainly comparable). The steady expansion of the size and contents of his botanical books parallels the explosion in the sheer number of specimens available throughout Europe for study, consumption, and commercial exploitation. In Turner's wake, the production of herbal books increased significantly in England, at least until the final decade of the century (Henrey 3). Turner was also significant as a cultural model because he strove to put English plants into print for the first time; in his 1568 *Herbal* he claims to describe only species "whereof is no mention made neither of ye old Grecianes nor Latines" (ii). This is not to say, however, that his work consisted only of work with words; we know also of his extensive travels, his plant-hunting expeditions, and his cultivation of gardens both in England and abroad. Without this contact with actual plants, he could not have written the books he did; but the fact remains that no extra-textual technique, formula, or invention lets Turner's botanizing stand outside his writings. His legacy therefore provides a testament to the central place of texts in the botanical work being done in his lifetime.

Turner's work embodies some important ways in which plants and texts interacted as objects of study and tools for study in this period. Scientific activity itself was still first and foremost a textual activity and only secondarily observational or experimental. As Ann Blair has noted, "Reading, compiling and commenting on texts long constituted one of the central practices of natural philosophy, from antiquity down to at least the late seventeenth century. From Pliny ... to the professors of philosophy in seventeenth-century universities, to do natural philosophy was in large part to gather, sort and critique ... texts" ("Annotating" 69). Turner was no revolutionary in this regard; the literary basis of his study of flora has been described in similar terms: "Turner ... shared to the full the literary and archaeological interests of renaissance scholarship. He examined and collated his authorities; he scrutinised his species in light of them; he laboured with nomenclature and synonyms: it might seem that catalogues were his chief concern" (Raven 134). Turner himself confirms this characterization when he classifies one of his early botanical books as nothing more than a catalogue of books he still aimed to write: in the preface to his 1548 *Names*, he refers to the publication as "a litle boke, which is no more but a table or regestre of suche bokes as I intende by the grace of God to set furth here after" (147).

Books might seem, then, to outweigh plants in the balance of Turner's botanizing; but no catalogue can stand alone as its maker's chief concern. A catalogue is necessarily metonymic, pointing beyond itself. The catalogue as a genre aims to render a collection more comprehensible by turning a non-textual object (like a plant or a book) into a text. But the conversion of a non-textual object into a catalogued text is not an end but a means to lead the catalogue's user back again to the catalogued objects. Humanist botany was, similarly, not just a matter of comparing texts with other texts but of "collating classical botanical texts with nature" itself in a manner that imaginatively integrated text and plant (Reeds 27). Consequently, even scholars "whose original motivation was philological, found themselves developing an independent interest in the local flora and fauna" (Peter Harrison 78). The botanical books of William Turner, whose avowed interest was not in philology but in flora, demonstrate the particularly privileged place of textuality in the period's understanding of plants, and the fact that Turner's botanical books were so devoted to the linguistic acts of naming and translating confirms the dependence of *herbis* upon *verbis* in the sixteenth century.

Biographical Contexts

Turner was born around 1508 in Morpeth, Northumberland, apparently the son of a tanner. With the patronage of Lord Thomas Wentworth he attended Cambridge from 1526 to 1533, gaining both a B.A. and M.A. While at university he was already undertaking botanical studies; at the same time, under the local influence of Ridley and Latimer, he also actively embraced church reform, well beyond the king's desired limits (Whitney Jones 9). In 1538 he published his first small book on botany, the *Libellus de res herbaria.* Around 1540 he was arrested and perhaps imprisoned, possibly for unlicensed preaching, for publishing heretical work, or for marrying despite having taken a deacon's vow of chastity, this latter being a newly-minuted capital crime (Carlson). Owing in some way to the ecclesiastical turmoil of late Henrician England, Turner went abroad at the start of the 1540s and used his years in Italy to earn a medical degree. He botanized while abroad in a variety of ecosystems across Europe, including modern-day France, Germany, Belgium, Switzerland, and Italy (Stearn, "Introduction" 7–8). He also continued to publish inflammatory Protestant tracts which were "eagerly read by his countrymen" (Jackson, "A Life" 16–17); they were also ordered burnt after a 1546 proclamation condemning Turner among other "divers leude and euil disposed persons [who] haue taken occasion to utter and sow abrode, by Bokes imprinted in thenglish tongue, sundry pernicious and detestable errours and heresies" (England and Wales). Six of Turner's publications were named, including one with a title (*The Spiritual Nosegay*) in which his spiritual and his botanical interests commingled, emphasizing the interpenetration of his concerns with the books of nature and scripture (Townsend 567). Only a year after the ban, however, Turner returned to a safer England under Edward VI. He appears to have been rewarded for his efforts at church reform by this more ardently Protestant regime when he found preferment as assistant chaplain and head physician to Edward Seymour, Duke of Somerset, under whose auspices Turner published his first English-language botanical book, the *Names of Herbes*, in 1548. Seymour was then Lord Protector of the young king and therefore, however briefly, the most powerful man in England;[1] such patronage gained Turner acceptance as a physician and visibility as a botanist among the elite.

---

1 1547 was the year in which Seymour was made Baron, Duke of Somerset, Earl Marshal, Treasurer of the Exchequer, Lord High Steward of England for the Coronation, Protector of the Realm, and Protector of the King's Person.

According to Turner's biographer, to combine the posts of physician and priest was not unusual (Whitney Jones 97). This occupational crossover suggests another basis for the herbalist metaphors in religious books of the period, such as *The Divine Herball* by Thomas Adams, five sermons meditating on botanical metaphors. Similarly, the *Herbal for the Bible* translated by Thomas Newton in 1587 declares itself on its title page to be "a plaine and familiar exposition of such Similitudes, Parables, and Metaphors, both in the olde Testament and the Newe, as are borrowed and taken from Herbs, Plants, Trees, Fruits and Simples," in which the literary explication is said to derive from "obseruations of their Vertues, qualities, natures, properties, operations, and effects"—observations of a sort that might lead to scientific publications in the later seventeenth century, but meanwhile served theological ends. Turner's colleague at Syon, the chaplain Thomas Becon, uses similar rhetoric to suggest a fusion of botanical and theological expertise in a devotional book, *The Flour of Godly Praiers* (1550), that elaborates on its pharmaceutical powers in the dedication:

> It is a flower, I graunte. Notwithstandynge suche a flower as if it be ryghtelye vsed, is of synguler vertue and myghtye in operacyon. No euyl ayer can hurte where the sauoure of thys flower commeth. Yea the deuil the world & the flesh can not abide the ayer of thys flower, so mighte is the spiritual operaction therof. The flower geueth a smel in the stretes to the soule of the faythefull, as Cimamone and Balme, that hathe so good a sauoure, yea a sweete odoure doth it gyue as it wer mirre of the best.

Turner, as mentioned, used similarly metaphorical frames to obscure the line between his spiritual and his medical work.

Turner almost certainly had the supervision and use of some of the gardens at Syon House (Chapman and Tweddle 9). A medieval abbey dissolved in 1539 and converted into a private house in 1547, Syon's renovation and secularization coincided with Turner's arrival. Monasteries had long been the sites of the most extensive and socially significant form of English horticulture until the middle of the sixteenth century, when—partly due to the state's redistribution of such ecclesiastical properties among the aristocracy and gentry—gardening became a much more widespread secular status symbol in England (Raven 192). The last significant period of ornamental and elite horticulture in England had been during the thirteenth century and had ended with the plague and the wars that followed in the fifteenth century. By the mid-sixteenth century, relative peace and increased trade abroad, as well as the boom in domestic renovation and construction, led to a resurgence in the practice and prestige of recreational gardening (Goody 150; Campbell-Culver 69, Peter Harrison 67–9). William Harrison attests to this horticultural renaissance in his 1587 *Description of England*:

> If you look into our gardens annexed to our houses, how wonderfully is their beauty increased, not only with flowers ... but also with rare and medicinable herbs sought up in the land within these forty years ... [I]n comparison of this present the ancient gardens were but dunghills ... (265)

Evidence from Turner's own hand suggests that Syon was an early English exemplar of a garden of rare specimens; for instance, in his *Names of Herbes* Turner notes that "Ascyron is not very common in England, howe be it I sawe it thys last yere in Syon parck" (159), that "Cypresses growe in great plentie in my Lordes graces gardine at Syon" (174), and that Ephemerum "groweth plenteously in germany but not in England that ever I coulde see, savynge in my Lordes gardine at Syon" (177). Precisely such rarities were favoured by aristocratic collectors. Turner's particular period of residence was extremely opportune: in the late 1540s—a brief period of "comparative prosperity and good harvests" (Loach 101)—Syon was a model of courtly culture, while its gardens served as templates for those participating in the new horticulture. Turner and his botanical work were thus connected to the very centre of English power.

Unfortunately for Turner, it was a quickly shifting centre; almost as soon as he and his work came into favour, they fell out again. Only a year after he dedicated to his patron, in 1551, the first volume of his projected masterpiece—an encyclopedic herbal in English that was only completed two decades later—Seymour was executed. After his patron's downfall, Turner struggled for another position, and finally succeeded in becoming the dean of Wells cathedral. But his two terms as dean (1551–53 and 1560–64) were interrupted by the reign of the Catholic Mary Tudor, when the life and works of the well-known author of *The huntyng & fyndyng out of the Romishe fox* were once again threatened. His books were again ordered destroyed in 1555 (William Thomas Stearn 7) and Turner once more fled to Europe to make his living in Germany as a doctor. But back in England under Elizabeth, he was apparently welcomed by the new Protestant regime. He was invited to preach a sermon at the prestigious site of Paul's Cross in 1559 before a large crowd including the city's mayor and aldermen (Jackson, "A Life" 21; Whitney Jones 37). Nor were his prospects diminished by the rapid rise to power of his former colleague from the Somerset household, William Cecil. Turner was reinstated at Wells, from which post he brought out the second volume of his herbal in 1562. Yet instead of mellowing with age or adverse experience, Turner's advocacy of ecclesiastical reform became only more rigorous. As early as 1559, his radicalism was alarming his bishop, who complained in a letter to Cecil of Turner's "indiscreet behaviour in the pulpit": among other offences, "he condemneth utterly all bishops, and calleth them white coats, tippet gentlemen, with other words of reproach much more unseemly, and asketh who gave them authority more over me than I over them" (qtd. in Whitney Jones 41). As his insults make clear, Turner was a vocal participant in the vestiarian controversy over clerical dress, then at its height. In 1565 Archbishop Parker wrote to Cecil that he had been "informed yesterday that Turner of Wells hath enjoined a common adulterer to do his open penance in a square priest's cap .... You of the Council know what ye have to do" (qtd. in Whitney Jones 41). Turner's lack of caution led to his suspension around this time for nonconformity. He moved to London where he died in July of 1568, the same year that the third and last volume of his herbal was finally printed.

As this account suggests, despite his relative obscurity as a cultural figure today, Turner did not work in isolation in his lifetime. However sporadically, he had the attention of some of the most powerful men of his generation. Aside from Seymour, Turner also had the ear if not the patronage of a man as powerful as Cecil, "a keen collector of rare plants" (Henrey 39) who had been with Turner at Cambridge and worked for Seymour when Turner did (Raven 94).[2] As a botanist he was surrounded by a group of like-minded physicians, apothecaries, herbalists, and gardeners, both at home and abroad. By coincidence, a man named Thomas Gibson, a contemporary of Turner's from his hometown of Morpeth, also pursued a career as a herbalist, physician, and reformer as well as a printer (Raven 52), and thus acted as almost a mirror image of Turner in uniting the fields of plant culture and print culture—a likeness Turner would not have appreciated, since Gibson reprinted the *Grete Herball* in an edition that Turner condemned in the preface to his 1568 herbal as "full of cacographees" (ii^{v}). In the prologue to the first volume of his herbal, however, Turner names other English botanical colleagues with considerably more respect:

> There haue bene in England, and there are now also certain learned men: whych haue as muche knowledge in herbes, yea, and more than diuerse Italianes and Germanes, whyche haue set furth in prynte Herballes and bokes of simples. I mean of Doctor [John] Clement, Doctor [Thomas] Wendy and Doctor [George] Owen, Doctor [Edward] Wotton, & Maister Falconer. (Aii^{v})

Here Turner immediately sets the study of herbs not in the context of gardens or laboratories but of "prynte" and "bokes." Clement was a Greek scholar as well as a physician, and thus another example of the humanist basis of such herbalism. Like Turner's, Clement's pursuit of herbalism in England was interrupted by periods of exile caused by his theological commitments: while Turner fled Henry and Mary, Clement left England upon the accessions of Edward VI and Elizabeth.

The social status of those noted for their involvement in plant culture has historically loaned it a social respectability that contributors to other manual crafts, such as metallurgy or anatomy, have not always enjoyed; Turner's list shows that even in his time, botany could borrow prestige from its practitioners, many of whom were physicians, including royal ones (like Wendy and Owen). Early modern botanists also made a point of concocting a prestigious pedigree for their studies by dropping the names of

> ancient heroes, sages, poets, kings, and emperors who were wise in plant-lore—Achilles, Pythagoras, Democritus, Cicero, Virgil, Mithridates, Agamemnon, Nero, Solomon, and many, many others. With these distinguished precedents to guide them, Renaissance botanists could get their hands dirty grubbing at plants without risking their dignity and authority. (Reeds 28)

---

[2] Turner appeals to Cecil for patronage to carry on with his natural-historical work in the preface to *The Names of Herbes* and, twenty years later, dedicates his *New Boke of Wines* (1568) to Cecil, while several letters from Turner to Cecil discuss preferment. Jackson calls Cecil Turner's "staunch patron" ("A Life" 23), but whether these bids were successful is less clear.

Turner himself, in the "Prologe" to the 1551 herbal, cites

> kyng Salomon, [for] that he was sene in herbes shrubbes and trees, and so perfectly that he disputed wysely of them from the hyghest to the lowest .... Therfor wher as Salomon was commended for the knowledge of herbes, the same knowledge of herbes was expressed ynough comendacyon. (Aii^{r})

Solomon was a valuable case for the chaplain and future dean to cite, not only for his status as king but for his prominence and status as a biblical figure.

Aside from being respectable, the study of plants was also becoming relatively widespread in England, or so Turner claimed in the preface of his *Libellus* when he modestly suggested in his letter to his readers that "there are six hundred of us Englishmen who in this kind of learning would precede me (as the saying goes) on white horses" (translated by Raven 68). He goes on to say, however, that since none of these worthies had undertaken to publish his knowledge of plants, he felt obliged to do so: botany demanded a place in English books. In putting plants into books, moreover, Turner was clearly confident of an informed audience. As well as addressing his fellow specialists, he addressed his botanical books to a broader interested readership. In his own research he consulted frequently with ordinary people, and specifically with women, who might well be better-versed than university-educated physicians in herbal matters owing to the convention of household medicine being women's work (Whitney Jones 59). Turner thus wrote his vernacular herbals with an eye to non-specialists as well; of his 1551 volume, he wrote that he "dowt[ed] not but many both physicyons of the mean sorte, many surgiones and potecaries, and many of the common people, that will wysely and warely vse herbes with the counsel of the phisycyan, shall take very great profit and commodyte" from his book (Aiii^{r}).

Turner certainly found such a wide audience, at least for his antics and tactics as a church reformer, which lived on after his death in popular legends about the notoriously radical "Doctor Turner" (Whitney Jones 36). More flatteringly, the pioneering nature of Turner's botanizing was singled out for honours by his contemporaries: he was described as "the father of English Physicke" by William Harrison in his *Description of England* (289), and John Stow wrote that Turner "was the first that, by great labour and travel into Germany, Italy and other foreign parts, put forth any Herbal in English" (quoted by William Thomas Stearn 8). His reputation as a medical author was recognized by other publishing physicians in his lifetime: *Bulleyns Bulwarke* names him in a list of "men of credite" who "haue written of this most worthie Arte, of the hande crafte of *Chyrurgi*, or meanes thereunto" and writes of "the most worthy doctor Willia[m] Turner" that "his boke of Herbes, will alwaies grow grene, and neuer wither" ("A little Dialogue, between Sorenes, and Chyrugi," Fol.iiii). Turner's work was also reputed abroad; the botanist and bibliographer Conrad Gesner, for instance, testified to his renown throughout Europe as a natural historian (Whitney Jones 200). His work on plants was thus available even within his lifetime as an outstanding model for other herbalists, for writers, and even for the common people whose wisdom he consulted before incorporating it into the books that appeared under his own name.

Anthological Botany

The early publication dates of Turner's *Libellus de re herbaria novus* (1538) and his *Names of Herbes* (1548) have earned Turner the characterization as "the author of the first two printed books relating to British plants which have any claim to originality" (Stearn, Preface ix). H.M. Barlow goes further in his study of old English herbals, calling Turner "the only original English writer on the subject [of herbalism] in the sixteenth century" (24). Even his most recent editors describe Turner as "the first Englishman to make an original contribution in his own language to the scientific study of plants and their use in medicine" (Chapman and Tweddle 16). In this way his work is implicitly contrasted with that of his medieval predecessors, and even with slightly earlier sixteenth-century works like Banckes' herbal, a book conventionally named for its printer instead of its author and one which is more of a translation and compilation than a recognizably "authored" work. The attributions of originality to Turner stand out all the more when set in comparison to the other most noted English herbalist of the sixteenth century, John Gerard, whose herbal is usually remembered as an example of flagrant plagiarism.

Turner thus remains renowned largely for being a recognizable "single author" of the sort whose accomplishment modern thinkers often find it easier to credit. Turner himself bolsters later claims about his originality when he boasts, in the subtitle to the third part of his herbal, that he includes "herbes, trees, rootes, and fruytes, whereof is no mention made of Dioscorides, Galene, Plinye, and other olde Authores." In the preface to the same volume, he appeals to his personal experience of plants as the basis for his claim to originality:

> I wente into Italye and into diuerse partes of Germany / to knowe and se the herbes my selfe / and to knowe by practise their powers and workinge / not trustinge onlye to the olde herbe wiues and apothecaryes (as manye Physiciones haue done of late yeres), but in the mater of simples myne owne eyes and knowledge: wherefore I haue somthinge of myne owne to present and geue ... (iii)

In this passage, the insistent redundancy of certain phrases—"to knowe and se ... to knowe by practise" and by "myne owne eyes and knowledge ... somethinge of myne owne"—emphasizes Turner's perception of the importance of his individual and even bodily involvement in the creation of the knowledge that went into his herbal.

Yet Turner is also known, outside his published works, to have participated in botanical correspondence that involved the inclusion of plant materials in packets of letters. During his first period of exile, for instance, he sent samples of plants from Brabant to the London apothecaries John Rich and Hugh Morgan (Jackson, "A Life" 17). He exchanged seeds and specimens with well-known continental herbalists (Raven 116–17; McLean 213; Arber, *Herbals* 121); he sent a copy of the first part of his herbal to the leading naturalist Conrad Gesner, who also published, in 1558, a text on fish that Turner had sent him the year before (Chapman and Tweddle 20). Such correspondence reveals that botany was a highly collaborative enterprise, the nature of which is reflected in the often anthological composition of herbals. Even Turner's work, then, can be seen to participate in the anthological orientation of sixteenth-century textual culture.

Not only did Turner depend on others for his knowledge of the names and natures of herbs and their habitats; the production of his grand opus, the three-volume herbal, necessarily involved a variety of contributors, as can be seen from the address added by Turner's son Peter. Immediately after this address, the herbal has attached to it page upon page of printing errors compiled by Peter after his father's death. The younger Turner writes in "Peter Turner to the Reader":

> ... this my fathers Herball, which he after his long trauell, study and experience hath made, is now set forth to profit and pleasure his countriemen withall: and by the ouersight partly of the Printer, and partly of the Scribe that copied the booke out for the Printer, is now so commed forth that the Reader can not get any profite or commoditie thereof ... (Ai).

This explanation incidentally points out the highly anthological nature of the work. To supplement his father's extensive contributions, the herbal was copied by a scribe and brought out by a printer before being amended by the author's son, who does not minimize his own contribution: "I haue taken some paynes, and haue pervsed and red ouer this booke as my Father beganne, and haue conferred it with his owne hande copie, and haue so corrected it and amended it." He then immediately anticipates criticism of his editorial choices:

> But peraduenture some will saye, if I had minded to pleasure my Countriemen, I might haue done better if I had called in or stayed this print, and caused the booke to haue bene printed here anew again. In dede if I had done so, I shoulde haue eased the reader of this labour of correcting.

His refusal to stop the press has the consequence of yoking the reader to the team of individuals whose work contributes distinct parts of the final product. Peter Turner, after all, has only compiled a list of errors; for that list to become useful, countless readers must in turn correct those errors in their own copies. Peter Turner goes on to allude to his own potential and more substantive contribution "when the Herbal shall be printed againe (and if it please God to lend me lyfe and health, augmented and increased)." This did not occur.

Despite some defensive emphasis on his individual accomplishments, seen above, William Turner could also show an awareness of the anthological nature of his herbal in his description of the labour that he put into the book. Most often, when he refers to his work on title pages and in prefatory material, he uses the rhetoric of "gathering." For instance, the title page of his 1548 book describes it as *The Names of Herbes ... Gathered by William Turner*; he is thus credited, or credits himself, exclusively for his work as a collector. This is a common way of describing the task accomplished by those cited on title pages in this period: *An Hospitall for the Diseased*, for instance, is referred to as being "Gathered by T.C." on the title page (Cartwright); *The Garden of Health* was "Gathered by the long experience and industrie of William Langham, practitioner in phisicke"; *A book of cookrye Very necessary for all such as delight therin* was "Gathered by A.W." The verb choice is doubly appropriate in Turner's case, since in order to achieve what he does in this book, he first had to gather the plants themselves—physically, laboriously, over time and hill and dale—before he could go about gathering their names in the variety of languages that he cross-references and from the variety of books that he consulted. Gathering rosebuds was not the leisurely stuff of pastoral after all.

The broad cultural significance of "gathering" as a rhetorical act has been established by Mary Thomas Crane, according to whom gathering was understood in the sixteenth century as the first of two primary acts undertaken in the process of humanistic textual compilation. While Crane concentrates on the place of gathering and framing in relation to the genre of the commonplace book, she also considers these two discursive practices to be the basis of both the theory and practice of all reading, writing, and discourse in the period (3–4). English humanism, specifically, was "largely based on the collection, assimilation, and redeployment of textual fragments" (6). One consequence of these discursive practices was "a version of authorship that was collective instead of individualist," which in turn fostered a sense of "common ownership of texts and ideas" (4, 6). Crane notes a certain set of commonplaces *for* the act of commonplacing itself, tropes that recur "almost relentlessly throughout the period," among which may be found countless metaphors of "planting and harvesting, poison and cure" (9). The rhetoric of gathering, then, depended on language indebted to botanical practices, at the same time as books of botany like Turner's depended on a rhetoric of gathering.

Turner simply gathered the names in his 1548 book, but three years later the first volume of his herbal is described on its title page as *A new herball, wherein are conteyned the names of herbes ... gathered and made by Wylliam Turner*. The addition of a second verb to describe its author's labour is interesting in part because it suggests a distinction: "to gather" is one thing, while "to make" is something else. But where did Turner, or others of his time, draw the line between the two acts? The editor's function as collector was not clearly distinguished from that of the author as maker. Although authorial ownership rights in copy were not yet established, some sense of authorial ownership and of the concomitant concepts of plagiarism and piracy certainly existed in Turner's lifetime. Despite the existence of a botanical and a broader humanistic culture that encouraged a commonplacing approach to texts, authorial possessiveness and litigation also found expression in the period in relation to printed botanical books. The controversy surrounding John Gerard's herbal provides one case. An earlier example is the apparently successful lawsuit undertaken by one German printer against another in 1533 in relation to the latter's use of woodcuts from the former's herbals (Eamon 110). William Eamon has shown that "[s]cientific and technical books were especially vulnerable to literary piracy," primarily because "plagiarism was especially difficult to prove in a work that ostensibly described nature," as opposed to works explicitly expressing a person's ideas (110). The widely-held conception of the divinely-authored Book of Nature might have encouraged the sense that no single person could own the knowledge extracted from natural things.

The prefatory rhetoric from *The Grete Herball* appears to allude to this sense of herbal knowledge as God-given and therefore common to all: "brotherly loue compelleth me to wryte thrugh y[e] gyftes of the holy gost shewynge and enformynge how man may be holpen with grene herbes of the gardyn and wedys of y[e] feldys as well as by costly receptes of the potycarys prepayred" (ii[r]).

In the preface to his 1568 herbal Turner acknowledges the potential for accusations of plagiarism to be made against his work—indeed, he suggests that the charge had already been made, persistently and by more than one accuser, but denies its substance. He acknowledges that

> some of them still saye, seynge that I graunte that I haue gathered this booke of so manye writers, that I offre unto you an heape of other mennis laboures and nothing of myne owne, and that I goo about to make me frendes with other mennis trauayles. (i)

In his reply, Turner partakes of the same botanical metaphor used by more typically literary authors and editors of anthological gardens of verse in defending their tendency to gather material from other writers and texts:

> if the honye that bees gather out of manye floures of herbes, shrubbes and trees that are growing in other mennis meadowes, feldes, and closes[,] may iustelye be called the bees honye: and Plinies booke de naturalis historia maye be called his booke although he haue gathered it oute of so manye good writers whom he vouchsaueth to name in the beginninge of his worke: So maye I call it what I haue learned and gathered of manye good authores not without great laboure and payne my booke and namelye because I handled no one Author.

It is perhaps in equal parts appropriate and ironic that Turner should use a botanical commonplace *about* commonplaces to defend what he characterizes as an example of the commonplacing habit in his botanical work. The analogy of the compiler and the bee had a long and distinguished history that has been thoroughly traced by Ann Moss in the context of the printed commonplace book (12ff; see also Havens, 13–16 and Lechner 138–42). Almost paradoxically, Turner rests the defence of his work as his *own* on the fact that it is compiled from so very *many* authors. By his logic, a little plagiarism is a dangerous thing, but a lot is authorship.

Turner's Reforms: Naming and Translating

In *A midsommer nights dreame*, Theseus describes the activity of the archetypal poet in the act of writing poetry:

> ... as
> Imagination bodies forth the formes of things
> Vnknowne: the Poets penne tunes them to shapes
> And giues to ayery nothing, a locall habitation,
> And a name. (G2[v]–G3[r])

In the last few lines he might as well have been describing the work of the botanist's pen in the sixteenth century. Turner portrays his work in this way in the preface to his *Names of Herbes*:

> Thys litle boke co[n]teineth the names of the moste part of herbes, that all auncient authours write of both in Greke, Lattin, Englishe, Duche and Frenche[;] I haue set to also the names whiche be co[m]monly used of the poticaries and co[m]mon herbaries. I haue tolde also the degrees of so many herbes as Galene the chiefe Doctour of al phisicians hath written of, and because men should not thynke that I write of it that I neuer sawe, and the Poticaries shoulde be excuselesse when as the ryghte herbes are required of the[m], I haue shewed in what places of Englande, Germany & Italy the herbes growe. (Aiii^{r})

The vast majority of Turner's two earliest herbal works consist of giving a local habitation and a name, if not several, to the plants catalogued therein.

The centrality of naming, in particular, to the work that Turner undertakes in his botanical books is apparent even in the names of the books themselves. The title of his first, *Libellus de re herbaria novus,* or "A new little book on herbal things," is (like so many titles on early printed books) self-conscious about the objective status of what it names: the book's very bookishness is emphasized from the first word. The same principle of attending to the material nature of the thing to be named is often at work when Turner deals with plants for which he can find no English name and is compelled instead to coin his own. The title of Turner's 1548 work, *The Names of Herbes*, signals the prominence of naming (as opposed to medicinal functions or habitats) in the book. Finally, in the title of the first volume of his herbal—*A New Herball, wherein are conteyned the names of Herbes ... with the properties, degrees and naturall places of the same*—Turner once again emphasizes his work with names and subordinates other botanical features as markedly secondary subjects.

Translation—the linguistic way of providing something with a "local habitation"—also self-consciously suffuses Turner's work with words. His very first publications were translations of controversial theological works; that these were banned in 1546 shows what could be at stake in the act of translation in his time. Moreover, the *Names* of 1548 is essentially an expanded vernacular translation of the earlier *Libellus* (William Thomas Stearn 6); the 1548 book, in turn, can be considered "a condensed first draft of the Herbal which ultimately replaced it" (Ravens 98). The *Libellus* itself consists of a list of 144 plant names, which are shown in translation among three languages: Greek, Latin, and English. In *The Names of Herbes*, Turner declares a more ambitious multilingual program on the title page: he now translates among as many as six different languages and even professional dialects: "Greke, Latin, Englishe, Duche & Frenche wyth the commune names that Herbaries and Apotecaries use." In the second part of his herbal, he occasionally adds another vernacular (Italian) into the mix. The extent of Turner's intellectual concern with translation may be inferred from a passage in that volume of 1562:

> I wold wish that they that fynde any old Greke examples or copies of old authores and intend to translate them, that they shuld as well set out and cause to be printed the Greke texts as theyr own translationes: for so myght men the better examin theyr translationes, and the studiouse youthe by comparyng of them together myght profit much more in the greke tong and practicioners myght be more bold to work accordyng to it that they have translated. (100–101)

Raven notes that Turner "followed out his own exhortations and compared versions with originals" (59). A devoted humanist, Turner also seizes every opportunity to educate the young in the classical languages.

Turner's work with names was closely conjoined with his work as a translator: "very many English names which are effectively translations of the Latin are directly owed to Turner" (Whitney Jones 68). This fact draws attention to Turner's fundamental assumption: that plants must have English names. Why could not he and his fellow countrymen simply use whatever Latin or Greek, German or French, Cambridgeshire or Northumberland name had most currency? Why add to the babble of Babel by adding to the word-hoard of another vernacular? Turner frequently prefers to coin an entirely new word that qualifies in his eyes as English. It could be that Turner's assumption points in the direction of a growing linguistic nationalism (Richard Foster Jones 68). The humanistic and pan-European concern for collecting and translating textual fragments was particularly pressing for the English, whose language was considered "impoverished and ineloquent, lacking an adequate number of words to express the concepts found in ancient texts"; collections of commonplaces, in particular, "provided an important way to enrich the vernacular" (Crane 14). The same concern was addressed by Turner when, in looking over lists of classical and continental plant names, he found gaps where English matches should be. Linguistic nationalism may explain why Turner is not generally content to continue using the Latin or Greek name, or a contemporary German one, for that matter.[3]

Translation had its broadest Christian context in relation to the confounding of tongues after the destruction of the Tower of Babel, prior to which the "whole earth was of one language" (Gen. 11:1), the language of Adam in Eden.[4] With the consequent confusion of tongues, language became conventional and "signs merely arbitrary representations of the things they signified. No longer could the nature of

---

[3] The matter is complicated, however, by the fact that he occasionally does take up a name from another language as a suitable one for English users. For instance, in *Names* he writes of Anchusa that "it is called in fre[n]ch orchanet, it may be named in englishe wilde Buglos or orchanet, as the french men do." If a French word will suit in this case, why not in all? Turner himself does not address such questions in general or explain his rationale in individual cases.

[4] The stories of Eden and of Babel were at this time, as perhaps now, "among the best known" Old Testament stories and were "alluded to frequently" in the sixteenth and seventeenth centuries specifically in order "to frame accounts of the progress of knowledge" (Bennett and Mandelbrote 7).

things be read from their names: the loss of the language of Adam entailed the loss of the knowledge of Adam" (Peter Harrison 225). If that original state could not be wholly recovered, it was thought that it could at least be re-collected in fragments by gathering together the world's many vernaculars, through translation, so that they could communicate among themselves. But this broad Christian context for translation was somewhat constricted by the more controversial connotations of translation during the Reformation. Turner's translations involved forging bonds across linguistic boundaries when theological translation was a highly fraught issue all across Europe, and his first two printed botanical books bracket a period when this was a matter of particular urgency in England. Early reformers of Turner's generation concentrated on arguments and acts in favour of translating sacred scripture from sacerdotal Latin into every European vernacular so that God's words would be available to the priesthood of all believers. A theological translation edited by Turner in 1548 emphasizes this aspect of church reform on its title page, where the book is described as "not onlye for curates [and] yong studentes in diuinitie: but also for al christen men and women what soeuer age they be" (Spangenburg). Turner was in his first year at Cambridge, studying under Ridley and Latimer, when Tyndale's revolutionary translation of the New Testament from the original Greek was smuggled into England only four years after Luther's German translation of the same. The Bishop of London confiscated and burned as many copies of Tyndale's work as possible; anyone in possession of a copy could also be burned. Turner fully embraced church reform in the course of his studies at Cambridge, where Pembroke College in particular was a hotbed for such changes (Creighton 16–17). He emerged as a graduate and an illegal itinerant preacher at about the same time, in 1536, when Tyndale was burned at the stake for his translations. In 1538, the king was refreshing his suppression of such literature with a proclamation that created press licensing, stifled debate on transubstantiation, exiled Anabaptists, and deprived married clergy like Turner of their positions (Clegg 26; Carlson). Yet in the same year, Miles Coverdale's Englished Bible (based in part on that of the recently executed Tyndale) was ordered by the king to be used in every church. Turner's first botanical work was conceived and produced in a highly unstable atmosphere in which producing and especially translating the wrong text at the wrong time could be fatal.

Not just translating itself, but reading certain works translated into the vernacular could be equally dangerous; in an Act of 1543, the reading of the Bible in English was forbidden to any woman not noble or gentle, as well as to "artificers, apprentices, journeymen, serving-men of the rank of yeomen and under, husbandmen and labourers" (Bennett 27). Turner publicly criticized this act the next year in "An addycyon to the Reader" appended to the second edition of his *Huntyng and fyndyng out of the Romyshe foxe*:

> After that I had fy[n]ysshed thys boke, it was tolde me that the bysshopes had made an Acte, that none but ge[n]tle men and gentle wemen myght reade the Scrypture, and certayne ryche men. But I wyll aske the autours of that Acte, whether they suffer the gentelles and the ryche to rede the Scrypture for theyr

> soules helthe, or for theyr pastyme. If they suffer them to rede it for theyr sowles helth, dyed not Chryste as well for craftes men and poore men as for gentle men and ryche men, and wolde not Christ that the poore labouryng men shulde haue wherwith they myght comforte theyr sowles as well as ryche men and gentle men?

Turner construes the reading of Scripture in relation to the "soules helthe," thus drawing together the spiritual and herbal genres in which he wrote. This book of Turner's was named among those condemned by the Act of 1546, the same year in which Rome reasserted its control over the interpretation of the Bible at the Council of Trent (Peter Harrison 94). Turner addressed the practical nature of the problem of the ecclesiastical aversion to the vernacular in *The huntyng of the romyshe vuolfe*, in which he wrote that priests teaching in Latin were like sentries at the Scottish border who alerted the townspeople with an incomprehensible "veniunt scoti in Latine / and would not say in English / the Scottes come": if, in consequence,

> the people woste not what the watchman sayde / and so were sodeinly taken of the Scotes / and were for the most parte killed / were he not a murderer / and were hee not worthy to be hanged for his laboure? Euen so / the parsone or byshop / that readeth the scripture / wherein are conteyned all wordes of warning of all men from all perilles in a strange tongue, that the people vnderstandeth not.

As this analogy suggests, from perspectives both practical and nationalistic, Turner valued translation in the realm of secular scripture very much as in the sacred. That Turner was equally concerned with the need to translate medical knowledge out of Latin and into the vernacular may be inferred from his book on *Spirituall Physik*, in which he writes out a prescription in Latin only to follow it immediately with an English translation and a jab at those who would refuse to provide one: "ye Popyshe prestes rede ouer ye sicke p[er]sons heades, Christes aboue named medicines i[n] latin, as though the bare redynge of the physicians byll in a straunge tonge, shulde helpe the syck men, when as they understande neuer one worde thereof" (63v). In this instance Turner himself connects the translation of sacred and secular texts, and specifically medical ones.

Under Edward VI the tide turned in favour of the vernacular translation of all things sacred. In 1548, the same year as the publication of the *Names of Herbes*, a royal proclamation added to the Latin mass a layman's communion service in English; the Book of Common Prayer was prepared in the same year, and when it was given statutory authority only a year later, the service was rendered wholly in English (Loach 50–51). Yet even as sacred translation was increasingly accepted and encouraged in church matters, secular translation in another area was threatened by its own particular priesthood—one to which Turner himself belonged. A faction of academically-trained physicians fought the tide of affordable, available, vernacular books of medicine (Eamon 102–3); when Thomas Elyot published in 1534 his first edition of *The Castel of Helth*, for instance, it "provoked criticism from the doctors" simply because "it was written by a layman and in English" (Raven 42–4). The resistance of physicians to vernacular translation appears to have gained special strength in the 1540s owing to the changes in England's health care after the Dissolution. Ecclesiastical houses had been a long-standing source of health care, in particular for the poor; with their dissolution, most infirmaries, hospices, and city hospitals were also shut down, at least until civic authorities refounded them (Hoeniger, *Medicine* 25). While Turner would not have regretted monastic dissolution in principle, he was no disciple of Henry VIII, having suffered under his policies to the point of exile. As a deacon and later a physician, he would have understood the drastic social consequences in store. While monks, friars, and priests had for centuries tended both the ill and the herbs that made them well again, Henry's rapid appropriation of ecclesiastical properties resulted in a shortage of medical personnel. As a result, "Protestant priests and ministers received strong encouragement from local bishops to step into the breach. Several divines also contributed to the publications in English of useful practical medical information" (Hoeniger, *Medicine* 24). In pursuing his degree in medicine when he did, Turner qualified himself to participate in this movement among reformers. In 1548, when the consequences of the dissolution were widely felt, Turner broke faith with many of his fellow physicians when he published an expanded vernacular translation of his 1538 Latin herbal work. This and his subsequent botanical books can thus be interpreted as efforts to address the transformation of health care in the 1540s, which was itself related to church reform. Even Turner's first botanical publication, although in Latin and thus only aimed at the learned, attempted to establish English names for medical herbs; his subsequent books on medicinal plants were all in English and dealt increasingly with local flora which could be better put to use by his countrymen than plants that only grew in southern Europe and Asia Minor.

Turner anticipated criticism from fellow physicians in making his herbal works available to the secular equivalent of the priesthood of all believers, as becomes clear in his dedication to the 1551 volume of his herbal:

> [o]thers will thinke it unwysely done, and agaynst the honor of my art that I professe, and agaynst the common profit, to set out so muche knowledge of Physick in Englyshe, for now (say they) every man ... nay every old wyfe will presume not without the mordre of many, to practise Physick. (Aiii^{v})

He answers this charge sharply by asking "how many surgianes and apothecaries are there in England, which can understande Plini in latin or Galene and Dioscorides, where as they wryte ether in greke or translated into latin" (Aiii^{v}). He points out the absurdity of the opposition to the vernacular when classical authorities wrote in their own native tongues:

> If they gaue no occasyon unto every olde wyfe to practyse physike, then gyue I none. If they gave no occasion of murther: then gyue I none. [i]f they were no hynderers from the study of lyberall sciences, then I am no hynderer, wryting unto the English my countremen, an Englysh herball. (Aiii^{v})

He furthermore suggests that most physicians know no more of plants than grocers and women, an act of levelling that was simultaneously a blow to his fellow physicians and a bolstering of lay practitioners.

Women and grocers had similarly been bolstered by the state with an Act passed in 1542. The Act was designed to support the informal system of health care that had often, in the past, been officially as well as unofficially suppressed and that had of necessity surged up in response to the disappearance of ecclesiastical hospitals. The state no doubt recognized the social utility of deeming that

> from hensforthe, it shalbe leful [lawful] to everye p[er]sone being the King['s] subject having knowledge and experience of the nature of herbs ro[o]tes and waters ... to practyse use and minystre [them] ... according to theyre cooning experience and knowledge. (Great Britain 906)

This act was remarkable for overriding acts passed earlier in the century that had restricted the legal practice of medicine to formally educated and licensed practitioners. The express endorsement of amateur herbalism came in response to the fact that

> the Companie and Felowship Surgeons of London, mynding oonlie theyre owne lucres, and nothing the profite or ease of the diseased or patient, have sued troubled and vexed divers honest p[er]sones as well as men and woomen, whome God hathe endued with the knowledge of nature kinde and operac[i]on of certeyne herbes ro[o]tes and waters ... and yet the saide p[er]sones have not takin any thing for theyre peynes and cooning, but have mynistred the same to the poore people oonelie for neighbourhode and Goddes sake and of pitie and charytie. (Great Britain 906)

The act places herbalism explicitly in the context of Christian charity. The appearance of Turner's *Names of Herbes* in 1548 is more understandable when interpreted as a response to and a participation in this aspect of medicine. Medicine was undergoing a reformation in tandem with that of the church, and vernacular translation occupied the same salient place in each.

The idea that Turner's Protestant theological work could have had a bearing on his botanizing may seem far-fetched at first glance, but the two subjects were not nearly as disparate then as now. Turner himself observed and even forged links between his work as a reformer and as a medicinal herbalist, as can be seen in the way he exploits, in his theological works, the metaphoric potential of botanical material. As mentioned, several titles attributed to Turner bear evidence of this potential, including his lost *Spirituall Nosegay* and his *New Booke of Spirituall Physik for Dyeuerse Diseases of the Nobilitie and Gentlemen of Englande.* The latter is identified on its title page as being authored by "William Turner doctor of Physik," as if to drive home the book's guiding metaphor; Turner was also an ordained deacon, but in this theological work he emphasizes not his ecclesiastical but his medical credentials. Another of his medicinally-inflected reformist tracts, *A Preseruatiue, or Triacle, Agaynst the Poyson of Pelagius*, is signed likewise. In *Spirituall Physik* Turner equates church reform and medicine when he refers to the way in which Henry VIII "toke in ha[n]de to reforme & heale some of hys clergie" (58^{r}) in the midst of his own diagnosis of the various diseases of the nobility, listing their symptoms and prescribing cures.[5] A scholar of early English Protestantism claims that Turner's use of medical metaphors in religious polemic was entirely novel, although the treatment of sin as a disease and Christ as a physician were certainly ancient and enduring tropes (Pineas 387–90).[6] Turner's rhetorical choices in texts like these show his own sense of the close integration of his two careers.

Turner's earliest publications participated in church reform specifically by engaging in acts of translation of theological works written by European contemporaries. His first extant publication, the 1534 *Co[m]parison* mentioned above, was printed anonymously, but by its second printing in 1537 Turner was emboldened to assert that it was *Translated out of Latyn in Englysh by Wyliam Turner*. Evidence from a letter written by Turner to William Cecil in 1550 suggests that Turner went so far as to propose tackling the task of editing an English translation of the Bible; he proposes that he "will ... correct ye hole newe testament in englishe, and wryt a booke of ye causis of my corretio[n] & changing of the translatio[n]" (quoted by Jackson, "A Life" 19). That Turner felt the need to follow his edition of an English translation of the Bible with a second volume vindicating the "causis" of the first is an indication of how fraught was the matter of biblical translation, while the next sentence in the letter—"I will also finishe my great herball"—shows how closely intertwined were his two lines of work. Another issue of his *Co[m]parison* was printed in 1538, the same year as his first botanical book, and yet another in 1548, the same year as *Names of Herbes*. Turner's translating work in both botanical and scriptural fields thus went on simultaneously and responded to similar concerns.

A further parallel between Turner's work as a reformer and as a botanist can be found in his attitude to ancient authorities. In the first theological work that he translated—*A co[m]parison betwene the olde learnynge & the newe*—Turner stakes his loyalties not with "the newe" but "the olde"; his idea of ecclesiastical reform was of a return to an allegedly original and untainted version of Christianity.

---

5 For those with "the Romyshe pokkes" caught from "spirituall fornicacion," for instance, Turner advises abstinence from "the Romyshe wyne and from the leuen of the pharisees, whyche is counterfet holynes," and prescribes a diet of the bread and water of life (Christ), as well as a purgative of repentance prepared from ".vi. drames of the bytter medicines of ye turned sorcerers, and .iiii. vnces of the aboue named water of lyfe, mingel them together and warme them vpon the coles of .... popyshe bookes, well burnt into coles, and then when the drynke is warme, drynke it hole out" (78^v–80^v).

6 Tessa Watt notes that in a series of early seventeenth-century broadsides responding to an outbreak of plague, "remedies based on walnut kernels, treacle or endive water are complemented with a metaphorical recipe: 'First, fast and pray, and then take a quart of Repentance of *Ninive*, and put in two hand-fuls of Faith in the blood of Christ, with as much Hope and Charity as you can get, and put it into the vessell of a clean Conscience'" (229). She also cites *The kings medicines for the plague. Prescribed for the yeare 1604. by the whole colledge of physitians*, which "recommends a combination of 'bodily' and 'spirituall' medicines involving herbs, bramble leaves, vinegar, patience, prayer and hope" (229).

Like other church reformers, his "objective was the restoration of the purity of original Christian teaching by removal of the spurious papist accretions of recent centuries" (Whitney Jones 11). This task bears a remarkable resemblance to that of a humanistically-educated herbalist of the same period. In the following description of Turner's herbalism, we can recognize the Protestant reformer in the botanist:

> [H]e is critical, often severely so, of those moderns who ignore or misinterpret the authorities of the classical age. His purpose was not to break away from the past or to rebel against tradition but to reaffirm and defend a yet more ancient wisdom. To do so involved scholarly handling of documents and the business of verifying their findings by precise definition of their content. (Raven 57)

Peter Harrison also notes the resemblance between scriptural and botanical reform: just as the Reformation "was an attempt to reconstruct Christian religion from its origins .... natural history, during its constructive phase [in the sixteenth century] at least, was similarly the reconstruction or restoration of a past discipline"—one with its own canon of literary authorities (97).

Turner's sense that the secular and sacred scriptures had shared stakes was by no means idiosyncratic. Peter Harrison writes that the Protestant emphasis "on the historicity of the biblical text, and on the circumstances of its human authors, had the inevitable consequence of reducing the gap between scripture and other writings" (124). Turner was thus not alone among his generation in connecting botanical and theological work; many early modern botanists were active in the reformed church. Otto Brunfels, Leonhart Fuchs, Jerome Bock, Carolus Clusius, Valerius Cordus and his father Euricius, Conrad Gesner, and both Bauhin brothers were Protestant, while Andrea Cesalpino, Ulisse Aldrovandi, and Garcia da Orta "were each at some time in bad odour with the Catholic Church on suspicion of heresy" (Morton 153). Bock was also a Lutheran pastor, while Brunfels (like Turner) wrote Protestant pamphlets, and both Fuchs and Cordus taught at the new Lutheran universities (Reeds 13). The *Reformation der Apotecken* (1536) by Brunfels was "part of an effort to reform pharmaceutical practice along the lines of the contemporary religious reform. Just as the Church had drifted away from its true scriptural principles, Brunfels asserted, so had medicine and pharmacy" (Eamon 98). Many of these herbalists were also, like Turner, pioneers in publishing their herbal works in their respective vernaculars. To some extent, a turn to the Book of Nature could be an attempt to circumvent the corruption that had crept into the Book of Scriptures:

> The book of God's works was older than the Bible whose interpretation caused so much controversy, and the hope was that by a common endeavour to observe nature and exchange information, scientists might recover the shattered unity of Christendom. (Prest 38–9)

The very leaders of the reformation, Martin Luther and Philip Melanchthon, "made clear not only their own personal tastes for gardening and botany, but also the theology that made such pleasures suitable for Christians" (Reeds 13). Luther, moreover, felt that Adamic nomenclature was an utterly perfect one: "because of the excellence of his nature, he views all the animals and thus arrives at such a knowledge of their nature that he can give each one a suitable name that harmonizes with its nature" (qtd. in Peter Harrison 249). The quest to restore such a perfect nomenclature was thus well aligned with the larger quest for church reform.

What's in a Name? Truth, Falsity, and Fiction

Nomenclature may be less obviously at stake than translation in effecting church reformation or in Christianity more broadly; but its significance in Turner's religion is as elementary as the act of naming itself. By the sixteenth century, to name and to christen were all but synonymous terms and tasks. In the first chapter of the first book of the Bible, immediately after creating various aspects of the universe, God names them: Night, Day, Heaven, Earth (Gen. 1:4–10). In the second chapter God turns the task of naming over to the first human; after forming the beasts and birds, God "brought them unto Adam to see what he would call them" (2:19). As a result, Adam's first act and his contribution to creation is to name "every beast of the field, and every fowl of the air" (2:19), and God accepts Adam's word as authoritative: "whatsoever Adam called every living creature, that was the name thereof" (Gen. 2:19). Adam appears to complete his task comprehensively with respect to the world's fauna, giving names "to all cattle, and to the fowl of the air, and to every beast of the field; but for Adam there was not found an help meet for him" (2:20). The semi-colon here signals an abrupt cessation in Adam's naming, however, and after taking a nap and waking up with a wife, he never apparently returns to the task, which means he never gets around to naming any plants. By Turner's time the imperative to finish the job was felt urgently once more. Turner's own herborizing forays into botanical things thus needed to be fulfilled, almost typologically, in textual form. By turning to the very basic task of correlating the corresponding names for plants among the Babel of European languages, ancient and modern and regional, Turner's work can be seen to participate in a contemporary response to the Christian imperative, increasingly felt and variously fulfilled in his time, to restore the garden of Eden on earth.

Aside from such spiritual purposes, though, names had more grounded uses ranging from the medicinal to the commercial. Clear and correct botanical names were fundamental to any advance in botanical studies, and according to Turner, they were lacking in his lifetime; his *Names of Herbes* is full of such phrases as "I haue not hearde the name of it in englishe" and "it hath no name in englishe that I knowe" (8, 10). Moreover, those names that did exist were not always well-known. As he notes in the preface to his 1568 herbal, it was not just the unlettered herb-wife who was unfamiliar with the names of plants; rather, it was the physicians themselves, from whom Turner complains that he

> could learne never one Greke, nether latin nor English name, even amongest the Phisicions[,] of anye herbe or tre, suche was the ignorance in simples at

> that tyme and as yet there was no Englishe Herbal but one, al full of vnlearned cacographees and falselye naming of herbes. (ii^r^)

Both the importance of correct nomenclature to Turner, and his insistence that medicine's bookishness should not prevent it from being a manual art, are apparent in the way he rails, shortly after this passage, against his contemporaries in the field: "howe can he be a good artificer that nether knoweth the names of hys toles nether the toles themselves when he seeth [them?]" (iii^r^).

The varieties of work that went into Turner's botanical naming may be surveyed in one of his lengthier entries, which explains some of the problems he and other herbalists of the era encountered as well as the solutions he found acceptable. The following is a translation of his entry for Narcissus in his Latin *Libellus*:

> For a long time among us Narcissus has concealed itself under foreign names. There was no herb that gave me more trouble. For after I had seen its picture and outline and studied them as carefully as I could, no one could be found to show me the plant, much less tell me its English name. At last when I was taking a holiday in Norfolk a little girl hardly seven years old met me as I was walking along the road; she was carrying in her right hand a bunch of white flowers; as soon as I saw them I thought to myself Those are Narcissi—for the description of them was still fresh in my mind; and I begged some of them from her. But when I enquired the name no reply was forthcoming. So I asked the folk who lived in the neighbouring cottages and villages, what was the name of the plant: they all answered that it was called 'laus tibi': I could get no other name from them. But when I got home I learnt that asphodel was called by many people 'laus tibi'. Then a little old man whose name is Guarinus Asshe, a canon of Barnwell Priory and well-skilled in herbalism, told me this plant was called French Gillyflower (gelofer). We must use that name until a better is found. (translated in Raven 63)

This anecdote reveals how Turner conceived of his task as an interpreter of the Book of Nature. He appears to have worked not as a sole author but as an editor collecting the multitude of names assigned by many authorities over time. The first sentence shows his editorial work as an act of reform and revelation, stripping away the disguising "foreign names" that concealed English plants, much as ecclesiastical reforms aimed to strip away traditions alien to the true church. Yet Turner's position is not as simple as that of someone tearing off a disguise, for he cannot locate the object whose disguise he wants to remove. He must seek both plant and name; neither thing nor word is at hand, and yet he has a dogged faith in both. When he spots a child carrying the very thing, moreover, he still cannot learn its name. His next tactic is to consult the child's elders "in the neighbouring villages," presumably on the grounds that a plant's ecological niche should correspond to a linguistic niche; thing and word are expected to grow together symbiotically. And indeed, the villagers present him with a name: "'laus tibi'." Despite their unanimity, Turner's response indicates that he remains unsatisfied: "I could get no other name from them." He would prefer the ambiguity of many names, it seems, to this particular name, perhaps because, after persistently seeking the plant in England as well as a corresponding English name, the name he is finally given—and not by hide-bound academics but cottagers—is in *Latin.*

The religious connotations of "laus tibi" might further account for Turner's dissatisfaction with the name proffered without exception in the village where the plant grew. The phrase was used in Latin church services that may well have been objectionable to one of Turner's radical persuasion. A considerable number of plant names with papist connotations were eventually altered as a result of the dissatisfaction of Protestants like him. One of the aims of reformers of his generation was to eradicate idolatry, such as saint worship, wherever it might be found, including in the names of plants (Goody 156). St. John's Wort is one plant still known by its saint's name, but many more had similar names in English until the Reformation. Long before that, in a newly Christian Europe, plants had similarly been renamed as part of the conversion of the continent: "Just as churches were built on sites of pagan worship, so herbs and flowers were literally christened" (Goody 121). Flowers like the ladyslipper were once known as *Our* Lady's this and that; similarly, Sweet William was once Sweet *Saint* William.[7] Simply changing the name of the plants "stripped [them] of their religious significance" (Goody 156), adding another adumbration of the unexpected significance of early modern botanical names in their broader cultural context.

When Turner returns home from his field expedition for the English name of Narcissus, he is finally able to dismiss the village name on the grounds that, as he then learned, "asphodel was called by many people 'laus tibi'." He began with neither name nor plant, and proceeded to the slightly better position of a plant with no name, then to a plant with an unsatisfactory name; back home again, he has reached the considerably more confusing position of having two plants known by the same name. But Turner's story has a happy ending: he adopts the name given, not by a village full of people familiar with the plant and unanimous about its name, but by a single church authority who "told me this plant was called French Gillyflower." Yet Turner is still not entirely at ease, as his postscript indicates: "We must use that name until a better is found." One is left to wonder what would qualify as a better name, and where exactly it would be found, since Turner leaves

[7] Some extent of the changes may be indicated in the list compiled by Goody: Black Bryony was Rosaries; Alchemilla was Our Lady's Mantle; Greater Budweed was Our Lady's Nightcap; Ribbon Grass was Our Lady's Ribands or Garters; Meadowsweet or Bridewort was Our Lady's Girdle; Dodder or Strangleweed was Our Lady's Laces; Bird's-foot Trefol was Our Lady's Slipper; Cuckoo-flowers or Milkmaids were Our Lady's Smocks; Harebell was Our Lady's Thimble; Canterbury Bells or Foxgloves were Our Lady's Gloves; Kidney Vetch or Lambs Toes were Our Lady's Fingers; Wild Orchids were Our Lady's Tresses; Cowslip was Our Lady's Keys; Thrift was Our Lady's Cushion; Portulaca was Our Lady's Purse; Primula was Our Lady's Candlestick; Solomon's Seal was Our Lady's Seal; Clematis was Virgin's Bower; Ground Ivy was Herb of the Madonna; Mint was Herbae Sanctae Mariae; Parsley was Our Lady's Vine; Lungwort was Our Lady's Tears; Lily was the Madonna Lily (156–7).

the principles of his nomenclature unstated, although it is clear that he is working with certain rules of his own, however implicitly and inconsistently.

As the example above suggests, one special problem that Turner encountered when naming and translating was linguistic variation within England. When regional names abounded and no firm standard version of English was established, his task was even trickier than it might have been. Perhaps as a result, Turner's evaluation of regional and other dialectical variants of plant names is by no means consistent. He does not, for instance, regularly judge one form of English to be better than another. He can write that "Oxys is called in English Allelua, Cockowes meate, and wod sorel," and pronounce no verdict among the variants (*Names* 200). Under Amarantus, he writes with equanimity of one kind that it "is called here in Englande of some purple veluet floure, of other flouramore" (*Names* 153); similarly, he notes without comment that Arum is called "in english Cuckopintell, Wake Robin, or Rampe" (*Names* 158–9). He seems to see the virtue, for the purposes of identification, of a wholly inclusive approach to nomenclature. Yet he does frequently distinguish between "englishe" and other dialects specific to named regions. Turner also sometimes suggests that different names are used by different classes of writer; under Gingidium, which he suggests is called "Cheruel" in English, he adds, "Cheruel is called of some wryters Cherephillum, & of the co[m]mune sorte Cerefolium" (*Names* 182). Here, as above, he does not seem to judge the usage of the "com[m]une sorte" to be in error. Turner shows a similar lack of bias against the botanical dialect of the apothecaries, who were often disregarded by physicians like him; Turner, in contrast, makes a specific point of including their terminology in his *Names*, announcing the fact on his title page. He even judges in their favour when he says of one plant that "the poticaries" call it "Lingua ceruinam. To whose iudgemente I rather assent, then to Ruellius & Fuchsius" (*Names* 204). Considering that the physicians Jean Ruel and Leonhart Fuchs were two of the most prestigious botanists of the sixteenth century, this is a bold statement.

Turner does not consistently judge any community to be the most correct; for instance, in one entry he writes simply that "Carex is called in englishe a Sege ... it is called in Northumberla[n]d Shearegrasse because it cutteth mennes ha[n]des that touche it" (*Names* 165–6). In this entry, "englishe" and local terms are set side by side, with their contiguity acting as an apparent declaration of their equality. His etymological explanation of the Northumbrian name could even be read as a defence of the fitness of the regional term over the national one; yet the very fact that the term used in Northumberland is not considered "englishe" may, indirectly, assert the authority of the first name over the second. That local names are not and cannot simply be adopted, however, as English names—even in the absence of a pre-existing English term—is suggested by the entry on Cetunculus, where Turner concocts an apparently novel name for a plant which already had a regional name: "It may be called in englishe Chafweede, it is called in Yorke shyre cudweed" (*Names* 167). The close resemblance of these terms—varying by just a syllable—only adds to the apparent insult that a plant's name in Yorkshire (one only included *after* his own coining) should not be considered "englishe."

Elsewhere, too, Turner refuses to take up a local name and instead invents one of his own: "Orobanche is so rare an herbe in Englande, that I neuer sawe it in al Englande, but in Northumberlande, where as it was called newe chappel floure. It may be of his propertie called Chokeweede, because it destroyeth and choketh the herbes that it tyeth and claspeth wyth his roote" (*Names* 200). In this case, Turner's aversion to the Northumbrian name may result from strict theological views in relation to the idolatrous use of flowers in chapels (Goody 156–7). These and other examples show that Turner certainly did not feel obliged to accept or promulgate the usage of a local name in his quest to set down English flora in writing.

Turner's idiosyncratic approach to nomenclature is what helps to tilt the verdict in favour of his being the first "original" writer on plants in England in the sixteenth century: he coined many names since accepted and perpetuated as common names: Loosestrife, Goatsbeard, Hawkweed, Larch, Wintergreen, Kidney bean, and Soapwort, among many others (William Thomas Stearn 4). Turner's willingness to fill a void can seem presumptuous, but the details of various entries show that Turner did not take plant names lightly, as becomes evident, for instance, in his concentration on the distinction between true and false names. In his entry in *Names of Herbes* on Heliotropium, for instance, Turner writes that "[t]hey are foully dec[e]yued and shamefully deceyue other whiche holde in their wrytynges that our Marigold is Heliotropium Dioscoridis." Such fierce diction adds an unexpected moral aspect to the apparently simple task of naming. Turner's frequent references to "false deceyuers" sound not unlike the complaints of the scriptural translator Miles Coverdale, in the prologue to his English bible, about "the gloses of our sophisticall doctours." It may seem odd to think in terms of the categories of truth and falsity when it comes to the linguistic material of names; the theory of an arbitrary relation between signified and signifier is so thoroughly entrenched as to make anything else hard to imagine. But in Turner's time, it would have been equally odd (at least for Turner) *not* to think of names as either true or false. Certainly, "[t]he particular denominations which Adam bestowed upon the animals were thought to have been part of a natural language in which words were not arbitrary tokens of the things they represented, but the expression of the true nature of things" (Peter Harrison 249); and it was this Adamic language which many reform-minded naturalists aimed to restore. Since, while some "writers suggest that Adam's knowlege enabled him to give the creatures fit names," others "stressed the converse: that from the names, would come knowledge" (Peter Harrison 250), names could be conceived of as not merely correct but positively true, bearing truth beyond themselves. For Turner a simple name could contain truth or falsehood, categories of not only practical but spiritual significance.

The deep cultural roots of a spiritual context of proper naming may be gathered from a tale told by John Gerard in his herbal. Gerard, who lacked Turner's Protestant commitments, also tends to lack his righteous indignation about errors in naming, over which he is often content to pass without comment. However, in this particular tale, retold from Augustine in his "Epistle to Saint *Ierome*," he demonstrates just how seriously the matter of naming could be taken in a Judaeo-Christian context:

> ... on a time a certaine Bishop hauing occasion to intreat of this which is mentioned in the fourth chapter of *Ionas* his prophecie (in a collation or sermon, which he made in his cathedral church or place of assemblie) said, that this plant was called *Cucurbita*, a Gourde, because it increased vnto so great a quantitite in so short a space, or else (saith he) it is called *Hedera*. Vpon the noueltie and vntruth of this his doctrine, the people were greatly offended, and thereof suddenly arose a great tumult and hurly burly; so that the Bishop was inforced to go to the Iewes, to aske their iudgement as touching the name of this plant. And when he had receiued of them the true name, which was *Kikaijon*: he made his open recantation, and confessed his error, and was iustly accused for a falsifier of the holy scripture. (*Herball* 400)

Turner sometimes expresses similarly evangelical fervour about apparently fine distinctions among names more and less accurate. For instance, about a certain variety of Absinthium, or wormwood, he notes that "Some take thys herbe agaynst the trueth for pontike wormwod"; of another variety, he says that "they are farre deceiued that vse this for potike wormwod" (*Names* 149–50). When writing on the palm in his *Libellus*, his concern may be explained by the ecclesiastical context of the error; the result would, for him, be not merely a matter of a misidentification but of idolatry:

> On the Day of Palm-branches as they call it I have often heard priests saying 'Bless also these palm-branches' when I could see nothing but sallow boughs. What others saw, I know not. If they were not supplying us with palm boughs, they ought to change their petition and say 'Bless these sallow branches'. It is a lie to call a sallow a palm. (translated by Raven 68)

Even Turner's emphasis on his eye-witnessing, here and elsewhere—such as when he writes in the preface to *Names* that "men should not thynke that I write of it that I neuer sawe"—may be best interpreted less in terms of an emerging scientific method and more in terms of the Protestant emphasis on the authority of personal spiritual experience. The business of naming simples was thus more complex than what met the eye, and could reflect, equally, Turner's "zeal for both botanical and scriptural accuracy" (Whitney Jones 58) as much as his desire to avoid poisoning patients.

Turner's concern about false names can also, however, be understood in terms of their pragmatic effect, since he wrote his herbals in large part from a medical point of view. From this perspective, the correct name of a plant was that given to it by an ancient authority; it was correct because only that name allowed the herbalist to locate the best recorded data about the plant (Stearn, "Use" 2). A good part of Turner's motivation for writing a vernacular book like *The Names of Herbes* can be put down to a modern twist on such pragmatism:

> There seems to have been nothing in English to help [humanistically educated herbalists] to connect the traditional simples and herb-women's lore of the countryside with the plants prescribed as medicines by their classical authorities; and even the common herbs regularly used in their drugs were often wrongly identified or replaced by quite different species. (Raven 97–8)

This may be why, when Turner refers to errors of botanical identification, he most frequently represents the errors as "abuse": that is, the practical misuse of one plant in a prescription when another is called for, with concomitant consequences for the body in which the prescription was used. In the entry for Daphnoides, for instance, he writes that "some abuse the seede of it for coccognidio" (*Names* 176); of Piperitis, he writes "If thys herbe be not it that it is take[n] for, the yealowe seedes whiche oughte to be whyte do onely hynder" (*Names* 205). The complexity of plant identification and misidentification, and the extent of Turner's grasp of that complexity, are both evident in the entry on Myrica:

> The Poticaries of Colon before I gaue them warning vsed for thys [Myrica], the bowes of vghe, & the Poticaries of London vse nowe for thys quik tree, the scholemaisters in Englande haue of longe tyme called myrica[m] heath, or lyng, but so longe haue they bene deceyued al together. It may be called in englishe, Tamarik. (*Names* 196)

Accuracy of identification matters, in this context, only to the extent that the practical effects of misidentification can be harmful; for instance, in the entry for Elioselinum, Turner notes that "Some haue taken thys for Apio, but the errour was not very greate because they are lyke in strength" (*Names* 177). Sometimes, then, an error in name was significant only if it effected errors in use; the reverse could also be true, with errors in use sometimes resulting in false names. As Turner writes of Meum, "Some Poticaries in Anwerpe vse thys herbe for Peucedano and so they cal it" (*Names* 195). Another entry clarifies why certain plants were habitually conflated: "it that is take[n] here for Adiantum is trichomanes in Dioscorides. Howe be it ye may vse trichomanes for Adianto, for they are of lyke virtue" (*Names* 151). In this case, Turner's otherwise vociferous concern with the truth and falsity of names is considerably dampened. As long as plants were regarded primarily for their medicinal effects, there could be little motivation, even among truth-seekers like Turner, to establish a thoroughly distinguishing nomenclature among plants alike in "virtue," or medicinal power, since substitutions among those that had similar effects would not be dire.

For all his concern with the true identification of plants by their correct names, however, when Turner does not know what he considers the "englishe" name of a plant, he does not hesitate to make up one or more names on his own authority and with reference to a remarkable variety of criteria. For instance,

> Acanthium is called in greke Acanthion, it is named of some herbaries carduus asininus, I haue not hearde the name of it in englishe, but I thynke it may be called in englishe otethistle, because the seedes are lyke vnto rough otes, or gum thistle, or cotten thistle, because it is gummy and the leaues haue in the[m] a thynge lyke cotten, which appeareth when they are broke[n]. (*Names* 150)

Turner hardly seems to help matters by suggesting not one but three possible names for the plant, none of which he appears to prefer. Turner takes his cue for all three names from aspects of the material form of the plant, rather than translating the existing Greek or Latin name ("fool's thistle"). In this way Turner's approach to naming resembles that of Otto Brunfels: "He did not attach importance to choosing one Latin name and sticking to it; for instance, he calls the various buttercups indifferently *Pes Corvinus, Coronopus*, or *Galli Crus*—all three names being suggested by the resemblance of the leaf to the foot of a bird" (Arber, "Medieval Herbalism" 321). Only two entries later in his *Names*, though, Turner shows that he could have availed himself of a simple translation; when speaking of a type of Aconitum, he translates the Latin literally: it "is called Lycoctonum, & in englishe it maye be called wolfes bayne" (151). Further in the entry on "wolfes bayne," in discussing one of two types, Turner also suggests a name based on the German: "the one hath leaues lyke crowfote, and blewe floures lyke hodes, and it is called muench kappen in duche, and it maye be called in englishe mo[n]kes coule or blewe wolfsbaine" (151). In his first suggestion, Turner's translation alludes metaphorically, as does the original German, to the material form of the plant. However, Turner's second suggestion leads in a very different direction: it distinguishes the sub-type from the type by colour, a non-metaphorical reference to a material quality of the plant. The great variety of principles at work in just these three examples shows that Turner's conception of the truth and falsity of names—at least when it came to his own coinings—was neither explicit nor susceptible to simple analysis. Instead, he appears to avail himself, like a poet, of all the resources of language: classical and vernacular, descriptive and metaphorical.

In a markedly metaphorical instance of multiple name-coining, Turner notes that Cingulum "groweth in the sea, & is like a gyrdel, wherefore it maye be named in englishe, fysshers gyrdle or sea gyrdel, or sea belte" (*Names* 169). Without batting an eye, Turner once again sets down three separate neologisms for a single plant, never distinguishing a better or best among them; nor does he seem concerned that all three might, confusingly, enter circulation. Similarly, elsewhere he notes that Asplenum "may be called in englishe Citterach, or Scaleferne, or Fingerferne" (*Names* Biiii^r). That Turner multiplies his English epithets with such regularity might suggest that he did not have the temerity to envision his herbal as an authoritative and argument-ending manual. Instead, we might take as exemplary his attitude in the entry on Bacchar: "I wyll determine nothynge in thys matter tyl I haue sene further. Let lerned men examine and iudge" (*Names* 160). His own proliferation of neologisms certainly suggests that he was not consistently attempting to sort out a single systematic nomenclature from the already confusing miscellany of duplicated botanical terminology. Turner often only adds to that flood with his hodge-podge of multilingual terms, regional names, and coinings. The very fact that pinning down a one-to-one relation between plants and names was not Turner's top priority implies something unexpected about the epistemological circumstances in which he conducted his botanical studies, and the way in which those circumstances governed his aims. There could be no knowledge-advancing point to offering several new names for one specimen, as Turner repeatedly does. Yet the nature of the three English names Turner offers for Cingulum reveals something of his sense of what mattered most in the naming of plants.

All three names are variations on the two observations he includes about the plant: its habitat ("in the sea") and its visual and perhaps tactile resemblance to a girdle (an observation translated, as it were, from the Latin). His first proposed name extrapolates from the plant's habitat to imagine who else might inhabit the same area, and, moreover, to a possible user of the plant and a use for it: hence the first and possessive term, "fysshers gyrdle." In other words, to name a plant, Turner undertakes the imaginative act of inferring a human counterpart to the plant and a cultural adaptation of it. This shows his tendency, common among herbalists of the period, to conceive of plants primarily in terms of their utility to humanity, but it is also a concrete instance of the creative literary license taken by botanists like Turner. When Turner turns from "fysshers gyrdle" to "sea girdle," he appears to retreat from the imaginative act that had filled out the botanical habitat with a figure that made use of the plant; now he simply sets two terms side-by-side in a kind of denominational shorthand, without any clear relation between the nouns ("sea" and "gyrdel"). Without a possessive relation, no single image coalesces, as it did with "fysshers gyrdle." But Turner's mind may instead be moving from one imaginative name to one still more imaginative, since a possessive connection between the terms of the second name could be inferred from the first neologism. This time, though, a more abstract imaginative act is required: the sea must be personified for a belt to fit around it, and a less easily belted thing is hard to suggest. The almost paradoxical nature of the image conjured by his second name once again points to a remarkably imaginative spark. Still more interesting, because apparently less so, is Turner's turn from "sea gyrdel" to "sea belt." Here the only notable difference is the dropping of a syllable and the adoption of a synonymous term. That synonymy, which seems to make the third name the least worthy of interest, displays precisely where Turner's interests unexpectedly lay. His lack of commitment to a particular phrasing indicates that what we might think of as the concept *behind* the name mattered more to him than the wording itself. In other words, the imaginative act that connected the plant's appearance with another familiar thing, via metaphor, and that specified that metaphor by association either with something relatively congruous (a fisherman) or utterly incongruous (the sea), mattered more than any single linguistic expression. If so, this suggests a wholly different approach to naming than the simple word-thing relation we take for granted as natural or logical. Turner's casual multiple name-dropping may thus suggest less about the unorganized state of botany and more about the cognitive habits of the author and quite possibly his audience. For Turner, and perhaps for others of his time, a name appears to operate primarily not as a word but as an idea. This subordination of the linguistic and elevation of the imaginative aspect of naming sets the botany of Turner's day in close relation to poetry.

These examples from Turner's texts show that choices made about naming—the apparently simple act of linking word and thing, and perhaps the most basic linguistic act—can reveal a great deal of the namer's thinking, not only about words but about the world to which they were applied. By concentrating so self-consciously on meaningful acts of naming, and not in just one but in a Babel of European tongues, Turner promoted a revision of the contents of the natural world as an unfinished script. Moreover, in his acts of naming and translating, Turner contributed to work that would lead, in the next century, to attentive study of the structural, physiological, and ecological contexts of plants, none of which can be rendered orderly without a nomenclature, a systematization of the connections between words and things. Meanwhile, working without such a system, Turner created his own conventions in coining and translating a considerable number of plant names, some of which are still used in the vernacular today. An understanding of Turner's approach to botanical reformation of the sacred and secular scriptures therefore both requires and rewards attention to the linguistic and textual issues and innovations preserved in his publications.

Chapter 4
John Gerard's Uncommon Ground

Where Turner's early work provides an apt opening bracket for the burgeoning botanical culture of the second half of the sixteenth century, John Gerard's *Herball* provides a fitting close. Gerard was born around 1545, when Turner was probably working on his first English herbal; Gerard's own botanical publications appeared near the very end of the sixteenth century, in 1596, 1597, and 1599. After Gerard's herbal, no other significant work in the same genre appeared in England until 1633, when the book that appeared was only a much revised edition of Gerard.[1] The age of the herbal—as distinct from more specialized genres separately aimed at smaller audiences of gardeners, physicians, or scientific botanists—was coming to an end by the time Gerard wrote. Indeed, the bibliographer of early English botanical and horticultural writing concludes that between 1600 and 1650, relatively few such works were published, compared with the productivity of the surrounding periods (Henrey 213). Gerard's work therefore illuminates how verbal and herbal work had proceeded in the half-century between Turner's time and his own. Yet as with Turner, Gerard remains of interest less for the positive results of his herbalism than for his work with words. In Turner's case, taking such an interest exposed how his attitude to textuality was influenced by his work as a church reformer, and how the two basic verbal procedures of naming and translating formed the centre of his botanical work. In the case of Gerard, though, very different aspects of the bookish and textual cultures of his day are brought to the fore by his work with plants, including his problematic approach to authorship, his participation in an anthology-oriented and commonplacing book culture, and his poetic tactics and tendencies when putting botanical things into words.

As a historical figure, Gerard most vividly takes form when some of the biographical contexts of his botanical authorship are set alongside those of Turner. Both were trained and practising medical men: Turner a physician and Gerard a surgeon. Both were therefore primarily interested in the medical applications of plant study. Both worked in the immediate milieu of one of the nation's foremost leaders, William Cecil: Turner was at one time a member of the same household, and later sought his patronage, while Gerard worked directly for Cecil from around 1577 until Cecil's death in 1598, cultivating his gardens in London and at Theobalds Park in Hertfordshire and dedicating his herbal to him (Henrey 39). Both were well travelled: Turner journeyed throughout much of Germany and Italy, while Gerard

---

1 John Parkinson's *Paradisi in Sole*, sometimes referred to as a herbal, is more of horticultural handbook, despite the fact that Parkinson was an apothecary. Parkinson did publish a herbal (*Theatrum Botanicum*) in 1640.

likely travelled on a merchant ship to Norway, Denmark, Swevia, Poland, Estonia, and Moscow (Jeffers 18). Both unsuccessfully sought academic posts: in 1596, Gerard asked Cecil to recommend him as herbalist of Cambridge, which Cecil apparently did not do (Jeffers 41); in 1550, Turner fruitlessly applied to be president of Magdalen College (Chapman and Tweddle 10). Just as Turner produced what is often remembered as the first "original" herbal authored by an Englishman, Gerard is remembered for his own botanical first: his *Catalogus arborum* (1596), a list of plants growing in his garden that is remembered as the first garden catalogue printed in England (Jeffers 41). The catalogue parallels Turner's first botanical publication in several ways. Turner's *Libellus* was also primarily a list of plants, and both books are modest botanical débuts of not much more than a dozen pages each. The contents of Gerard's catalogue, like those of Turner's *Libellus*, appear in alphabetical order, a textual framework instead of any more inherently botanical one. Both books are primarily in Latin, the language most appropriate to work intended for an educated audience, and both were succeeded by larger vernacular versions (in Gerard's case, a bilingual edition in 1599). Gerard's modest catalogue was also shortly followed by his monumental herbal, just as Turner's first two brief publications of 1538 and 1548 played the part of pastoral antecedents to his epic three-volume encyclopedia published in later decades.

Parallels abound, yet Turner and Gerard were also very different men. Turner was a committed and vociferous church reformer, whose publications in that field rivalled—both in quantity and in impact—those he produced as a medical herbalist. Gerard shows no such commitment, confining his remarks on the divine to a few clichéd comments on the hand God had in producing some of the more marvelous botanical specimens. Turner had his M.A. and M.D., whereas Gerard, as a surgeon, did not require or receive any university training. While Turner's reputation was relatively fleeting among a popular English audience, Gerard's name was long remembered both by a popular readership (Hoeniger, *Medicine* 47) and—if only to revile it—by historians of botany.[2] Even in his own lifetime, Gerard appears to have attained greater and more lasting recognition than Turner in their mutual field of work, probably in part because Gerard's career was uninterrupted by the periods of exile which caused Turner to lose his place. Gerard became curator of the physic garden of the College of Physicians from its inception in 1586 and apparently continued with it until 1603; he succeeded in 1604 to the more prestigious position of Surgeon and Herbalist to James I, and was later elected master of the Barber-Surgeons company (Jeffers 30, 79–81). But while Turner remains reasonably well respected by historians as the proverbial "father of English botany,"[3] Gerard's reputation fell rapidly from grace.

---

[2] In a recent analysis, Rickman if anything politely understates the prevailing tenor of historians of science when she says that "most twentieth-century scholars agree that Gerard's *Herball* does not increase his merit or prestige"; she adds that he was strongly censured by both of the leading historians of English botany, Agnes Arber and Charles Raven, who "calls Gerard a rogue on two occasions" (34).

[3] Turner was apparently first identified as such by Richard Pulteney, in 1790, in his *Historical and Biographical Sketches of the Progress of Botany in England.* One of

The Garden Grows: From Catalogue to Herbal

Gerard's catalogue of the contents of his Holborn garden is alphabetized; the herbal printed just the next year is not. Gerard's decision to alphabetize would have affected the ways in which the catalogue could and could not be used. It was, for instance, emphatically not a map to the placement of individual plants in his garden. The textual order that shapes the list could only have related arbitrarily to the organization of plants in beds and pots, which are best arranged with respect to relative height and available light, material factors rather than something as ideational as the alphabet. The catalogue's lack of correspondence with anything but a textual order would mean that not just any interloper could use it to bridge the gap between text and plant while standing in the Holborn garden. Instead, the gardener himself remained the singular and necessary intermediary between these two substances. While the alphabet operates in one sense as a transparent public language, in the context of the catalogue it also operated as a private code to which the garden's gate-keeper alone held the key.

The botanical catalogue, as a genre, is a kind of hybrid of text and plant, in which names stand in metonymically for absent materials. The catalogue is thus a version of Gerard's garden translated into the medium of words. Such a reductive reproductive mode was far from beautiful and failed to be maximally useful, as noted above; it must, however, have had advantages, or else Gerard would not have troubled to assemble and print it. And indeed, Gerard's work with words ratcheted skyward the personal and public value of whatever work he did with plants. The advantages of creating a literary counterpart to a botanical collection were several, but foremost was the fact that Gerard's personal collection of plants could be made infinitely better known by being re-collected in textual form. The publicity that Gerard created for his horticultural work by transforming it into a text may have initially or apparently best benefitted a reading public with an interest in horticulture since, in a period before commercial nurseries, private gardens like his were often the best or only way to access rare and valuable specimens; but Gerard also certainly stood to benefit from the magnification of his standing as a botanical virtuoso. The catalogue in many ways seems a hasty compilation; printed only a year before his herbal, it may have been designed to establish with speed Gerard's standing as an authority on herbal matters. Yet he already had authority as a plantsman through the patronage of the most powerful man in England, the endorsement of the College of Physicians, and the possession of a notable garden of his own. Still, he made the decision to go into print, perhaps out of a conviction that bibliographic authority would best complement his already-established botanical authority. As in the nearly contemporary case of Ben Jonson, publication may have been the best way for Gerard to compensate for his lack of

---

Turner's earliest biographers, B. Daydon Jackson, followed suit in 1877 ("A Life" 15). Eleanour Sinclair Rohde, an early-twentieth-century writer on English herbals, noted that he was "commonly known as the father of English botany" (75).

academic credentials and show a broad public that he was up to the task of writing books as well as reading them.

The catalogue provided a means to broadcast or disseminate—two metaphors from the sowing of seed which today are bound up in the idea of communications media—the experience of Gerard's garden to a wider public than was possible without this translation of plants into words. The Holborn garden was fixed in the ground, not portable like a book. Like all known gardens at the time, it was almost certainly walled and thus emphatically private. The catalogue, on the other hand, was easily portable and could be "visited" simultaneously by any number of the reading public situated almost anywhere, whereas the experience of the garden as a whole, in its full living form, required on-site visitors. Only a textual version of the garden could act as a remotely approximate substitute for, or memoir of, a personal visit that took in the entire garden. That such was the response to catalogues like his may be inferred from a surviving response to what became known as Tradescant's Ark, a seventeenth-century collection of curiosities that included the very first English garden open to a paying public. The language of one visitor suggests that he considered the consultation of the garden's catalogue to be metaphorically equivalent to a personal experience or observation of the actual plants:

> In the museum of Mr. John Tradescant are the following things: first in the courtyard there lie two ribs of a whale, also a very ingenious little boat of bark; then in the garden all kinds of foreign plants, which are to be found in a special little book which Mr. Tradescant has had printed about them. (translated in Boesky 322)

The plants themselves, the writer imagines, "are to be found in a special little book," the garden catalogue; text and garden are imagined to reflect one another to the point of interchangeability. A historian of collecting elaborates on the parallel between experiencing an artifactual collection and a textual collection in this and other "literary analogues" to such museums: "Just as visitors toured the museum to witness tangible evidence of exploration and conquest, so readers perused the *Catalogue* .... As Tradescant's *Catalogue* exemplifies, the enterprise of compiling a catalogue parallels that of collecting, and both are reproduced in the mind of the visitor or the reader by experience" (Belk 162–3). The transference of the act of collection to the mind of the reader and the reproduction of the collector's labour by the reader are both suggested by the Latin terms—*legere* and *collegere*—for the two activities.

While the catalogue may be the more faithful textual translation of his garden, Gerard claims the herbal of 1597 enacts a similar role:

> bicause gardens are priuate, and many times finding an ignorant or a negligent successor, come soone to ruine, there be that haue sollicited me first by my pen, and after by the Presse, to make my labours common, and to free them from the danger whereunto a garden is subiect: wherein ... I was ouercome, and ... brought this historie, or report of the nature of plants to a iust volume ... (A2v)

It was a marked exaggeration for Gerard to represent the herbal as a reflection of his garden: the book details far more species than even he was able to cultivate. But Gerard blurs the distinction between his two books and their botanical counterparts when he carries the conceit of the book-as-garden as far as his proem, where he refers to the herbal's divisions into books and chapters as separate gardens and beds: "In three bookes, therefore, as in three gardens, all our plants are bestowed: sorted as neere as might be, in kindred and neighbourhood" (1). He extends the trope pedantically, taking a pleasure not unusual among his contemporaries in drawing out the analogy between the parts of a book and the parts of a garden.

Gerard concludes by offering to lead his readers through the gardens on the pages that follow: "hauing giuen thee a generall view of this garden, now with our friendly labors we will accompanie thee, and lead thee through a grasse plot .... Then by little and little conduct thee through most pleasant gardens, and other delightfull places" (1). Gerard frames himself as a gardener and guide in the herbal, the same roles he would have had in relation to those consulting his catalogue. These two very different books are likened not only by his construing the herbal as a representation of his garden but by his personally taking on the same crucial tasks in both books. The herbal itself is construed through his metaphorical language as an arrangement of delightful "places," a word that would likely remind contemporary readers of another bibliographic species: the commonplace book, comprising a different type of textual *topoi*.[4] The way Gerard frames himself as gardener-guide characterizes the herbal as a place that its author inhabits, a piece of real estate in which he is present throughout a given reading, offering his company and informed direction to each reader. Gerard's stylistic choices throughout the text repeatedly refresh the sense of his presence, since he does not shy away from the use of personal pronouns or anecdotes wrought from his own experience. Gerard thus presents his labour in creating the herbal as subject to continual renewal every time the book is opened. His horticultural work—far from ending with the cultivation of the Holborn cornucopia or even with its textualization in the herbal—continues at the behest of every reader who opens the book. As he writes, "with our friendly labors we will accompanie thee."

Gerard inscribes himself so pointedly in the pages of the herbal that, if the catalogue is the more accurate literary translation of his garden, the herbal can be read as a textual version of the gardener himself. This idea of Gerard's presence within the herbal makes more than metaphorical sense. All the hard-won knowledge about plants that would otherwise be contained only in his mind—a place, like the

---

[4] Beal concisely explains the origin and importance of the term: "The importance of what were known as 'commonplaces' was a tenet of Renaissance Humanist thinking and derived directly from classical teaching. For Aristotle, for instance, the proper treatment and classification of 'topics' was one of the fundamentals of the science of rhetoric .... Expounding Aristotle's rhetorical system later, Cicero defined a 'topic' as 'the place [or locus] of an argument,' and he said, 'Those arguments which can be transferred to many cases, we call common places'" (134–5).

garden, necessarily quite private and firmly *in situ* wherever his body might be—was translated into the herbal and thence disseminated in portable form to a large public. One of the prefatory poems alludes to the herbal's embodiment of Gerard's collected botanical learning. The surgeon Thomas Thorney begins his poem to his "Learned friend and louing brother in Art" as follows:

> Oft haue I heard, and oft haue reade
>   In bookes of learned lore,
> That Man, the name of *Little world*,
>   Or *Microcosmos* bore.
> And rightly sure; whose minde doth range
>   The circled world about,
> Whose head (a litle Globe) conceiues
>   Each worldly thing throughout.
> As this in all, so now in thee,
>   This thing appeereth trew
> By speciall note (sweete *Gerard*) heere
>   In this thine Herball new.
> Wherein (as in a glasse) we see
>   How thou thy minde hast bent,
> Thy bodie toyled, thy time bestowde,
>   And many a pound hast spent ... ($B2^{r-v}$)

Thorney's initial use of the microcosm trope mixes indiscriminately, a few lines later, with a mirroring trope. Both characterize the relation between Gerard and his herbal: one, as an analogical relation, the other as one of reflected identity. Thorney first suggests a ratio-like analogy *between* the analogy of man-as-microcosm and that of the herbal-as-Gerard: "As this in all, so now in thee, / This thing appeereth trew ... / In this thine Herball new." The herbal is a microcosm to Gerard's macrocosm; the herbal is also, as we learn a few lines later, a mirror to Gerard's "minde." Yet Thorney does not portray the herbal as a static reflection of a static object: instead, he suggests that the mirror-herbal reflects a moving pageant of Gerard's labour as herbalist, one depicting "[h]ow thou thy minde hast *bent*, / Thy bodie *toyled*, thy time *bestowde*, / And many a pound hast *spent* ..." (my emphasis). Thorney's verse lands hard on his deliberately varied verbs by their placement at the end of each phrase and line, each verb describing a different type of labour—the same "friendly labors" with which Gerard offers to "accompanie" his reader.

Thorney's verse is easily dismissed as mere unctuous bumph. After all, how can Gerard's labours be mirrored in his herbal? The herbal is neither a narrative of his travels as a plant hunter nor a record of his finds; it does not even, like the catalogue, mirror the garden he made. How else, then, might his labours—as gardener, botanical collector, and creator of the book—be mirrored in it? What comes to mind is to look to the labour that readers must engage in to profit from the book: reading and perhaps memorizing the words, connecting them with accompanying images, annotating the entries, distinguishing plants easily confused, and finally taking all of this arduously acquired knowledge back to the garden, field, or stillroom where it could come to some practical use. The book mirrors Gerard's labour whenever it is opened by a reader who mirrors Gerard by mimicking, as he picks his way through the text, the labour Gerard undertook in the field and study—though in reverse order, like any proper mirror image. The words with which the reader must work stand in, at least initially, for the plants handled by Gerard. Such continuity between writer and reader is embodied even in Thorney's clichéd phrase later in his poem: "we shall reape / The seede which thou hast sowne." In the herbal, the seeds are words, which the reader "reap[s]" first by reading and then by translating them back into the plants that they describe.

The Anthology of Nature: Gerard's Herbal Authorship

A little later in his poem, Thorney characterizes Gerard's achievement as one of "Preferring still the common good, / Neglecting still thine owne"; that is, he contrasts Gerard's private labours with the public benefit they yield. Gerard himself claimed that it was not by his garden but only "by my pen, and after by the Presse" that he could "make my labours common." With respect to Gerard's translation of his botanical collections into print, both men appeal to the same adjective, one with a host of connotations that Gerard and his associate may well play on. Once again, the disparate materials of book and plant are allied in a single word, since the promise of Gerard's herbal seems to be to unite the territorial idea of "the commons" and the textual idea of "the commonplace" by embedding his private property—both the contents of his garden and his related intellectual property—in a published book. "The commons" amounted to a form of national common ground: territory that was, historically, "undivided land belonging to the members of a local community as a whole" (*Oxford English Dictionary* 5a) and that was under increasing threat throughout the sixteenth century. Gerard may well have appealed to the popular critique of enclosures of the commons by construing the herbal—which he characterizes as a public version of his otherwise private garden—as a transformation of certain kinds of otherwise private land into a kind of commons. It is certainly true that the promise of a textual commons, open to all comers, could not have been remotely fulfilled in the case of a book as massive and heavily illustrated, and therefore expensive, as Gerard's herbal. But the promise nonetheless makes itself felt by Gerard and Thorney's reliance on a single word, one in which botanical and book cultures and concepts of ownership were equally embedded.

A second connotation of the adjective "common" was the commonplace, a piece of text construed as free from even such loose ideas of proprietary authorship as existed in Gerard's time, when "all texts formed a common storehouse of matter," at least theoretically (Crane 6). The analogy between herbal and literary commonplaces was certainly available earlier in the sixteenth century, with tenor and vehicle reversed: Luis of Granada proposed in his *Rhetorica ecclesiastica* "exploiting the dialectical places of argumentation like a pharmacist with his recipe for mixing medicines, the raw ingredients of which are herbs gathered in many locations .... to be stored in a commonplace-book ..." (Moss 203). Gerard's herbal writing also partakes regularly of what can be, perhaps generously, called a kind of commonplacing method. Even Gerard's aim to "make [his] labours common" was a herbalist's commonplace: Henry Lyte similarly defended his translation of the Dutch *Cruydeboeck* into English in his address to the reader: "I thinke it sufficient for any, whom reason may satisfie, by way of answeare to alleage this action and sententious position: *Bonum, quo communius, eo radius el præstantius*: a good thing the more common it is, the better it is" (iii). Gerard himself draws on and reproduces the words and work of many "authors" (832), as he refers to both his poetic and his botanical sources. His apparently generous act of making "[his] labours common," then, might also have been intended to compensate for his tendency to assume that the work of others was equally susceptible to the collecting of a scrap here and there. His generosity may, in other words, be a defense against the very charges of plagiarism with which his book was soon and enduringly plagued.

It was around the end of the sixteenth century that such ideas about a common stock of literary property began to undergo change, even if authorial property rights were not yet institutionalized. As Joseph Loewenstein observes, "the *law* of intellectual property" is not necessarily "a significant index of *cultural experience* of intellectual property," and "at any given cultural moment ... rival reifications of the cultural status of intellectual property may coexist" (21). He specifies the late sixteenth century as a time when "[t]he suite of practices and affects that we call authorship ... was thrown into a rich flux" (40). In its midst, what we now call Gerard's herbal appeared. Elizabeth Eisenstein, in her pioneering study of early modern print culture, alludes to the viability of an analogy between property rights in texts and those in territory when she refers to the way in which "a literary 'common' became subject to 'enclosure movements'" (120–21). Crane expands upon the same analogy in her study of English humanist commonplace conventions and their relation to early sixteenth-century consolidation of land ownership: the rhetorical strategy of 'gathering' commonplaces, she suggests, was "based on the liquidity of printed texts ... which rematerialized itself through the concept of content as 'matter,' and which 'reterritorialized' itself through the system of logical and rhetorical 'places'" (15). She argues that "[b]y inventing a less threatening form of capital to accumulate"—that of the textual commonplace—"humanists attempted to create for themselves a stable place in society between the landed nobility and the deterritorialized merchant class" (15).

Gerard himself alludes to plants being a comparably "less threatening form of capital" in his herbal's address to his readers, when he speaks of his work as a matter of "descrying of such a harmlesse treasure of herbes, trees, and plants, as the earth frankely without violence offereth unto our most necessarie uses." The vivid contrast with violent and harmful mined treasures makes clear that his argument is about plants as a form of capital. Botanical treasure was "harmlesse" because partaking of it did not, he implies, need to involve plundering any recognizable sort of property. Instead, Gerard frequently points out particular locales where medicinal weeds grew freely. One of the herbal's prefatory writers singles out this generous habit as one of Gerard's special virtues: "least the Reader should too often languish with frustrate desire, to finde some plant he readeth, of rare vertue, [Gerard] spareth not to tell (if himselfe haue seen it in England) in what wood, pasture or ditch the same may be seene and gathered." For instance, Gerard notes that "water Iuie ... is verie rare to finde; neuerthelesse I found it once in a ditch by Bermondsey house neer to London, and neuer else where" (680–81). His inclusion of the only known whereabouts of this rare plant turns that location into a different kind of common place, by publicizing rather than hoarding this potentially valuable information. Gerard is even more specific in pinning down where he found a certain kind of moss "in great abundaunce in a shadowie ditch vpon the left hand neere vnto the gate that leadeth from Hampsteed heath Toward Highgate" (1371). He confirms that he did not hesitate to shed light on this "shadowie ditch" in his next phrase, when he notes that he had already "shewed [the same place] vnto diuers expert chirurgions of London, in our wandering abroad for our further knowledge in simples" (1371). Their harvesting of the plant from this place did not threaten Gerard's ability to continue to do the same, any more than one author's deployment of a textual fragment deprived his fellows of it. Gerard could thus afford to make common the places in which he garnered the botanical form of his cultural capital.

One further connotation of the term "common" is present elsewhere in Gerard's own prose when he refers to Europe's first botanical garden as "the common garden at Padua" (1309). Gerard himself had been put in charge in 1586 of what might have been England's first institutional botanical garden, that of the College of Physicians (Henrey 39). Shortly before hiring Gerard, the College also undertook to compile a medical encyclopedia (Jeffers 29); the impulse to create the garden was thus likely generated by the preceding impulse to create a book. But the physicians did not manage to reap any paper-fruit from their garden until a generation later in 1618, when the first edition of *Pharmacopoeia Londinensis* was printed. This publication was, unlike Gerard's English herbal, entirely in Latin. It thereby converted what Gerard might have preferred to understand as a "common garden" into a relatively private place hospitable only to those who could read its Latin signposts. Gerard's decision to publish in English may have directly contravened the linguistic preferences of the College of Physicians as they eventually appeared in their own publication and as they were expressed throughout the sixteenth and seventeenth centuries. Their aversion to the vernacular was a part of their commitment to stamping out unlicensed practitioners; since its chartering in 1518, the College had eagerly prosecuted all sorts of traditional healers, a group that tended to be made up of those who could read, if anything, only their native tongue (Pelling and Webster 178, 183; Kremers 91). Fifty years before Gerard wrote, in his herbal of 1551, Turner anticipated accusations from fellow physicians that his vernacular herbal should lead to the "mordre of many" (Aiii) by enabling unlicensed practice. In light of Gerard's connection with the "common garden" of the physicians and his decision to scorn their linguistic elitism and scoop their project in his publication of an English herbal, one final connotation of "common" may be worth mentioning: that of the "vulgar tongue" (*OED* n.1, sense 12b).

The generally accepted account of Gerard's involvement in the production of what is known as his herbal begins in the 1580s, when the physician Robert Priest reputedly began a translation of a Dutch herbal by Rembert Dodoens, with his expenses paid by the printer John Norton (Jeffers 28–9). Priest died before the publication of his translation; as the manuscript is lost, no one knows what stage he had reached (Jeffers 93). Norton assigned the remainder of the job to Gerard. At some point—under the aegis of Norton, Priest, or Gerard cannot be known—the order of the plants was made over to reflect not the complex arrangement of Dodoens but that in a herbal by Matthias L'Obel. The images were borrowed by Norton from the Antwerp printer Christophe Plantin, and were those originally used by the herbalist Tabernaemontanus. A good portion of the work that Gerard apparently had to do involved matching the woodcuts to their textual counterparts; he is sometimes criticized for having done a singularly bad job of this. Thomas Johnson, editor of the 1633 edition of Gerard's herbal and so perhaps not his least biased critic, wrote that Gerard, "(as it seemes) hauing no great iudgement in them [the figures], frequently put one for another ... and oft times by this meanes so confounded all, that none could possibly haue set them right" ("To the Reader"). Partly due to these errors, Gerard gained and retained a remarkably bad reputation among botanical scholars. Inaccuracies aside, however, Gerard has more often been condemned for something amounting to plagiarism. Johnson is also the originator of this critique: he accuses Gerard of "making it a thing of heare-say, that Dr. Priest translated *Dodonaeus*" ("To the Reader").

Of more interest than this intractable dispute is the extent to which Gerard's work has been framed in terms of single authorship and originality in a period when an anthological approach to the construction of herbals was the norm. This was certainly true earlier in the sixteenth century when the anonymous English herbal known as the *Grete Herball* declared itself, in its preface, to have been "compiled / composed and auctorysed by dyuers & many noble doctours and expert maysters in medycynes / as Auicenna, Pandecta, Constantinus, Wilhelmus, Platearius, Rabbi moyses, Johannes mesue, Haly, Albertus, Bartholome & more other" (ii) and actually consisted, according to a modern analysis, of "a translation of a French original which is itself largely derived from the *Ortus Sanitatis*; and this again reproduces in a debased form material from *Das Buch der Nature* by Konrad von Megenberg, from Bartholomaeus and from Albertus Magnus" (Raven 62). Agnes Arber (who herself condemns Gerard for plagiarism) elsewhere notes the general tendency for herbal work of the period to be treated as common stock; ironically, she makes this point in relation to the difficulty of assessing the originality of the work of Dodoens, the herbalist whose work Gerard is supposed to have stolen:

> It is particularly difficult to appraise with any exactness the services which Dodoens rendered to botany. He and his two younger countrymen, de l'Ecluse and de l'Obel, freely imparted their observations to one another, and permitted the use of them, and also of their figures, in one another's books. (*Herbals* 83–4)

A similar example of the tendency to compile and anthologize herbal works may be found in *The garden of health conteyning the sundry rare and hidden vertues and properties of all kindes of simples and plants*, identified on its title page as having been "Gathered by the long experience and industrie of William Langham, practitioner in phisicke" in the year of 1597, the same year as Gerard's herbal;[5] Raven describes it as "an alphabetical series of plants" that "contains hardly a single original word" (45–6). Early modern herbals like these were books of flowers in both senses of the word: in their botanical content and in their florilegic form.

Gerard is held to a different standard than his contemporaries in part because he appears to deny working with Priest's manuscript. He refers only once to "Doctor Priest, one of our London Colledge, [who] hath (as I heard) translated the last edition of Dodonaeus, which meant to publish the same; but being preuented by death, his translation likewise perished" ("To the courteous and well-willing Readers"). Gerard's words tend to be contrasted with those of the physician Stephen Bredwell, who acknowledges in a prefatory letter the contributions of "the authors of this booke: to euerie one (no doubt) there is due a condigne measure"; among others, he singles out "D. Priest, [who] for his translation of so much as Dodonaeus, hath herby left a tombe for his honorable sepulture." Bredwell's metaphor—characterizing the translation as "a tombe," a lasting monument—is supposed to contradict Gerard's account that the translation "perished," and is usually assumed to be authoritative. It seems possible to reconcile these accounts by reading Gerard's words metaphorically and by keeping in mind the idea, explored earlier, of the herbal as an embodiment of Gerard's knowledge as a herbalist. In these terms, Gerard's assertion that the translation "perished" need not be read as a denial of his use of Priest as a source. Instead, if publishing was imagined as giving life to a book, he might simply indicate that the book that might have become known as Priest's herbal perished when Priest himself did before bringing it to print; it was reincarnated, however, as Gerard's herbal. Bredwell goes on to refer to the herbal as a "bookebirth ... brought foorth by Gerard," and suggests that in publishing it, he "brought foorth the fruit of it"—two metaphors that again attribute the allegedly living text to the one who laboured to deliver it to the public. At the same time, Bredwell foregrounds a kind of collaboration as the ideal model for such work:

> This bookebirth thus brought foorth by Gerard, as it is in forme and disposition faire and comly, ... so is it accomplished with surpassing varietie, to such spreading growth & strength of euerie lim, as that it may seeme some heroicall impe of illustrious race, able to draw the eies and expectation of euery man unto it.

While the "spreading growth & strength of euerie lim" likens the herbal's chapters to living branches, "imp" was a synonym for a botanical scion or graft, a shoot artfully attached from one to another growing plant in order to create a stronger hybrid. This is precisely the sort of hybridizing work that Gerard can best be described as doing in his herbal.

---

5 According to the ESTC, the printed date of 1579 is a misprint for 1597.

Gerard's characterization of Priest's work may be no whole-hearted endorsement, but he does acknowledge Priest's work and that of many others. In other words, Gerard does not portray himself as the sole author of his herbal, which makes accusations of plagiarism look a bit ridiculous. He instead portrays his work in this way:

> I haue here therefore set downe not onely the names of sundrie plants, but also their natures, their proportions and properties, their affects and effects, their increase and decrease, their flourishing and fading, their distinct varieties and seuerall qualities, as well of those which our owne countrie yeeldeth, as of others which I haue fetched further, or drawen out by perusing diuers Herbals, set foorth in other languages, wherein none of our countrie men hath to my knowledge taken any paines, since that excellent worke of Master Doctor *Turner*: after which time Master *Lyte* a worshipfull Gentleman, translated *Dodonaeus* out of French into English; and since that Doctor *Priest*, one of our London Colledge, hath (as I heard) translated the last edition of *Dodonaeus*, which meant to publish the same; but being preuented by death, his translation likewise perished: lastly, my selfe one of the least among many, haue presumed to set foorth vnto the view of the world, the first fruits of these mine owne labours. ("To the courteous and well-willing Readers")

In this passage, Gerard describes his tasks as threefold: he "set downe" factual information about plants, "fetched further" or "[drew] out by perusing diuers Herbals" plants which did not grow in England, and "set foorth" the herbal, or published it. None is necessarily the task of an originating modern-minded author; nor does all that setting, fetching, and drawing sound like the work of any airy early modern muse. It is not, moreover, the case that Gerard takes credit for the work of other herbalists without acknowledgment. He mentions by name the three English men known by him to have written on herbal matters and acknowledges himself as "one of the least among many." Throughout the rest of the herbal, too, Gerard frequently cites the work of other contemporary authors. He highlights moments when he quotes someone directly and frequently attributes ideas to sources even when he is not using their words. To give just one example:

> *Hirculus* is a plant very rare, which as yet I neuer sawe, notwithstanding we are greatly beholding to *Carolus Clusius*, the father of forren simples, who finding this plant among many bunches or handfuls of Spikenard, hath made it knowen vnto posterity, as he hath done many other rare plants[;] in translating of *Garcia* the Lusitanian phisition, he setteth it forth with a light description ... (920)

None of this sounds like the patter of a plagiarist. It thus seems more accurate to say that Gerard's herbal grew just as his garden grew: collaboratively, or anthologically. Gerard speaks frequently of those who contributed specimens to his private plot; for instance, he acknowledges the contributions of

> a worshipfull and learned gentleman, a diligent searcher of simples, & feruent louer of plants, master *Thomas Hesket*, who brought the plants thereof [of his village near Lancaster] vnto me for the increase of my garden. I receiued some

> plants thereof likewise from master *Thomas Edwards* Apothecarie in Excester, learned and skilfull in his profession, as also in the knowledge of plants, vnto whom I rest bounden as well for this plant, as also other rare and strange plants and seedes, especially for the plant of *Iuca*, which his seruant brought from the Indies ... (89)

In a few brief sentences, Gerard acknowledges three collaborators, the first two named and the last nameless, but still included. Later, he recounts a tale about "a kinde of Balme, or Balsame tree" from "the west Indies," of which

> [m]y selfe with diuers others, as namely Master *Nicholas Lete,* a worshipfull marchant of the citie of London; and also a most skilfull Apothecarie, Master *Iames Garret,* ... haue receiued seedes thereof from the right Honorable the Lord of Hunsdon, Lord high Chamberlaine of England, woorthie of triple Honor for his care in getting, as also for his curious keeping rare and strange things brought from the farthest parts of the world; which seedes we haue sowen in our gardens ... (1347)

Many leading herbalists were also among his personal acquaintances. He was in close correspondence with John Robin, herbalist to the French king, with whom he exchanged many specimens. Gerard was also well acquainted with the Flemish L'Obel, who lived in England after 1584 and who assisted in the publication of Gerard's herbal. He even corresponded with Rembert Dodoens himself (Jeffers 64).

The physician Stephen Bredwell also portrays Gerard's work as depending on a raft of prior authors, as if he saw Gerard primarily as an editor of earlier works. Bredwell characterizes the herbal as a kind of anthology of the work of "fiue men" in particular and puts especial emphasis on their plurality:

> What is he then that will denie his voice of gracious commendation, to the authors of this booke: to euerie one (no doubt) there is due a condigne measure. The first gatherers out of the Ancients, and augmentors by their owne paines, have alredie spread the odour of their good names, through all the lands of learned habitations.

In the margin next to the previous sentence is a list of these "first gatherers ... and augmentors": "Turnerus. Dodonaeus[.] Pena. L'Obelius. Tabernamontanus." Bredwell continues:

> D. *Priest,* for his translation of so much as *Dodonaeus*, hath hereby left a tombe for his honorable sepulture. Master *Gerard* coming last, but not the least, hath many waies accommodated the whole worke unto our English nation: for this historie of plants, as it is richly replenished by those fiue mens labours laid together, so yet could it full ill have wanted that new accession he hath made vnto it. Many things hath he nourished in his garden, and obserued in our English fieldes, that neuer came into their pens to write of. Againe, the greatest number of these plants, hauing neuer beene written of in the English toong, would haue wanted names for the vulgar sort to call them by: in which defect, he hath beene

> curiously careful, touching both old and new names to make supplie. And least the Reader should too often languish with frustrate desire, to finde some plant he readeth, of rare vertue, he spareth not to tell (if himselfe haue seene it in England) in what wood, pasture or ditch the same may be seene and gathered. Which when I thinke of, and therewithall remember, with what cheerefull alacritie, and resolute attendance he hath many yeeres tilde this ground, and now brought foorth the fruit of it, whether I should more commend his great diligence to attaine this skill, or his large benevolence in bestowing it on his countrie, I cannot easily determine.

Bredwell makes the collaborative nature of the enterprise clear, as well as the extent of Gerard's contributions. Bredwell would of course have known from experience that Gerard was no sole author, since Gerard attests frequently to Bredwell's assistance in supplying plants from his garden.

Like Bredwell, Gerard shows himself open to an anthological approach to the production of herbal books when he anticipates that his work will become fertile ground for later herbalists: he suggests in his dedication that "if there be no further fruit" of his labours than the first edition of the herbal, "yet this is of some vse, I haue minstred matter for riper wits, and men of deeper iudgement to polish: and to adde to my large additions where any thing is defectiue, that in time the worke may be perfect" (A3). He repeats his desire for others to continue where he leaves off elsewhere in the dedication to his "courteous and well-willing Readers": "may my blunt attempt serue as a whetstone to set an edge vpon some sharper wits, by whome I wish this my course discourse might be both fined and refined." He specifically invites readers to continue his unfinished labours in the book itself by filling in gaps with their own pens; he even claims to aid them in leaving space on his pages (or "bed"s) "where the diligent searcher of nature may, if he so please, place his learned obseruations" (2), while elsewhere he implies that he expected readers to finish his labours by colouring the herbal's woodcuts (153). But far from simply suggesting that his own edition may be modified by the annotations of individual readers, Gerard anticipates further editions; for instance, in his discussion of five sorts of Canterbury Bels, he speculates that "[t]he greatest sort which I haue set foorth not before described, shal rest and content it selfe with the name set downe in the title vntil some second writer shall adde thereto, or else referre it to a further consideration" (366). Gerard creates a sense of his text as provisional, one draft in an ongoing and broadly anthological process.

It is thus odd that Gerard is condemned for plagiarizing when he could be praised for compiling, a respectable enough habit in his day and particularly in his field. Even the rearranging of Dodoens's material by L'Obel's method has been interpreted in a sinister light as Gerard's attempt to disguise his theft; but even assuming the choice was made by Gerard instead of Priest or Norton, such an alteration could as easily constitute a creative leap: to take the textual content from one author and combine it with the preferable form of another. The result would be novel and possibly illuminate something that the two texts, each on its own, did not reveal. This kind of work was legitimated by the humanist paradigm of textuality which Crane has characterized as one of "gathering and framing," the two activities that she postulates as the basis of both the theory and practice of all reading, writing, and discourse in the period (3–4); "framing" is a perfect description of the rearrangement for which Gerard has so often been blamed. Crane further notes that "[t]his educational program in turn gave rise to a version of authorship that was collective instead of individualist" (4). In consequence, it was quite normal for early modern books to have many "authors" and to have such authorship embedded in the quotations that riddled the pages. Turner lists, after the table of contents of his 1568 herbal, 36 distinct sources. *Bulleins bulwarke of defence against all sicknesse* lists no fewer than 43 ancient and contemporary herbalists and describes them as "The Authours, Capitaines and Souldiours of this Bulwarke." Bullein specifies on the title page that another of his medical works is "collect[ed] out of many approued authours" (*A newe boke of phisicke*). Gerard also seems to have seen himself as a collector; on the title page he describes the herbal's material not as having been written but "Gathered by John Gerarde." He thus portrays himself as a kind of anthologist harvesting the flowers from previous herbals.

Whatever Gerard's debts to his predecessors, evidence of his original work in the herbal abounds. It may be found when Gerard insists upon examining others' assertions independently and coming to conclusions based on his own experience:

> The 18.kinde of *Tithymale* hath a rounde roote like a small Turnep, as euery author doth report: yet my selfe haue the same plant in my garden which doth greatly increase, of which I haue giuen diuers vnto my friends, whereby I haue often viewed the rootes, which do appeere vnto me somewhat tuberous, and therein nothing answering the descriptions which *Dioscorides*, *Pena*, and others haue expressed and set foorth. This argueth that either they were deceiued and described the same by hearesaie; or else the plant doth degenerate being brought from his natiue soile. (406)

He speaks often of things that others report "which I coulde neuer obserue, although I haue endeuored to finde out the truth of it" (614). He emphasizes "triall" and "experience" no less than Turner. Gerard was thus a distinctive author as well as an anthologist of others' herbal writings.

Botanical Poetics and the Rhetoric of Description

Just as Gerard's name is preserved and propagated on the title page of his herbal, the names of many other contemporary herbalists are preserved and propagated in the very plants that they studied: the familiar lobelia, for instance, is named after Gerard's sometime colleague and rival, Mathias L'Obel. Plants, like books, were susceptible to the concept of "authorship" and authorial entitlement as part of the sixteenth-century botanical renaissance. Gerard includes a paean to this particular aspect of botany in his address to the reader at the start of the herbal:

> Euerie greene Herbarist can make mention of the herbe *Lysimachia*, whose virtues were found out by King *Lysimachus*, and his virtues no lesse eternized

> in the selfsame plant than the name of [the artist] *Phydias*, queintly beaten into the shield of *Pallas* .... What should we speake of *Gentiana*, bearing still the cognizance of *Gentius*? Or of diuers other Herbes, taking their denomination of their Princely inuentors?

Gerard elsewhere confirms that it was for finding the herb that Gentius was so honoured: "*Gentius* king of Illyria was the first founder of this herbe, and the first that vsed it in medicine, for which cause it was called *Gentian* after his own name" (352). He cites these examples as part of his argument that herbalism can bring "euerlasting honors," not only to those who practise it but to "Princely" sorts who could afford to patronize it. The term "inuentor" is especially interesting in this light: with it, Gerard suggests that those who discover a plant or patronize its discovery have a role in its creation. One type of creation that we know plant discoverers undertook at this time, and that in spades, was the creation of botanical names. If naming plants was indeed imagined as a way of inventing them, it seems that much more reasonable that people making things with words—the poets and printers of the many contemporary gardens of verse—might liken their inventions to the vegetable realm, which was imagined to be brought into being by the application of words to things.

This particular connection between plants and books is more logical than it might at first look, since it was only by publishing their discoveries and their names for them—by converting their botanical work into textual work—that herbalists could assert their authorial entitlement to something about them. For instance, when Gerard describes two varieties of cowslip, which he labels as "*Auricula Vrsi ij. Clusis*" and "*Auricula Vrsi iiij. Clusis*," he explains their names by noting how "*Carolus Clusius* setteth foorth in the booke of his Pannonicke trauels ... [these two plants] which he hath founde in his trauell ouer the Alpes, and other mountains of Germanie and Heluetia" (641); the implication is that Clusius can claim the plants as his not simply because he found them but because he published his findings. The practice was legitimized by ancient precedents: "*Aesculapius* ... called it after his owne name *Asclepias*, or *AEsculapius* herbe, for that he was the first that wrote thereof" (751). Gerard never makes any such claim for himself, once again confounding claims of his self-aggrandizement as an author.[6] The closest he comes to taking such credit is in a name he assigns to a plant in honour of a fellow herbalist: in a chapter on Gumme Succorie, he names the second kind after Jean Robin, the French king's herbalist; Gerard explains, "I haue named [it] *Robinus* gum Succorie (for that he was the first that hath made any mention of a seco[n]d kind, which he sent me as a great dainty, as indeede I confesse it)" (225). John Gerard's naming the plant "Robinus" exemplifies how herbalists presented each other with verbal gifts in the same way as they exchanged herbal ones.

Gerard may not name any plants after himself, but he tells a tale of his naming that is an exemplary instance of the pointed exclusion of a discoverer's name. He describes a type of Panax which he claims he was the first to add to

6 He does discuss a plant called "herbe Gerard" (848), but makes no mention of any special association between himself and this plant.

the pharmacopoeia: "we haue added another sort, whose vertues we founde out by meanes of a husbandman, and for that cause haue named it *Panax Coloni*, or Clownes woort" (850). Gerard thus names the plant after the discoverer, but at the same time excludes the discoverer's personal name; the husbandman's involvement in the plant's discovery is memorialized only by the inclusion of a relatively derogatory term for his relatively low social status. Gerard attributes the discovery to himself ("we founde out"), and treats the husbandman as a mere medium (or "meanes") for his discovery. The tale Gerard goes on to tell suggests that professional jealousy might have been the reason he was not more generous to the husbandman; under the section on the plant's "vertues," we learn that

> [t]he leaues hereof stamped with *Axungia*, or Hogs grease, and applied vnto green wounds in maner of a pultis, doth heale them in such short time & in such absolute maner, that it is hard for any that hath not had the experience thereof to beleeue: for being in Kent about a Pacient, it chanced that a very poore man in mowing of Peason did cut his leg with the Sieth, wherin he made a wound to the bones, & withal very large & wide, & also with great effusion of bloud, the poore man crept vnto this herbe which he brused in his hands, & tied a great quantitie of it vnto the wound with a peece of his shirt, which presently stanched the bleeding and ceased the pain, insomuch that the poore man presently went to his daies worke againe, & so did from daie to daie, without retiring one day vntill he was perfectly hole, which was accomplished in a fewe daies by this herbe stamped with a little Hogsgreace, and so laid vpon in maner of a pultis, which did as it were glewe or soder the lips of the wounde together, and heale it according to the first intention (as we tearme it) that is without drawing or bringing the wounde to suppuration or matter, which was fully performed in seauen daies, that woulde haue required fortie daies with Balsam it selfe:

Gerard's spiteful side emerges in a postscript:

> I sawe the wounde, and offered to heale the same for charitie, which he refused, saying, that I could not heale it so well as himselfe; a clownish answer I confesse without thankes for my good will, whereupon I haue named it Clounes Woundwoort as aforesaide. Since which time my selfe haue cured many greeuous wounds, and some mortall, with the same herbe ... (852)

The story makes clear not only that the husbandman discovered a better cure than any previously found in Gerard's medicine case, but that he distrusted Gerard's medical authority, preferring his own. Gerard reasserts some of his own authority in his narrative by belabouring technical details and "tearme[s]" and proceeding to describe saving a gentleman and "a Shoemakers seruant" from some gory wounds, thus leaving the impression that he was the more heroic "discoverer" of the plant than the nameless husbandman. However ungenerous this instance may be, Gerard still shows a certain ethic of citation in acknowledging the discoverer at all, and in resisting the urge to name the plant after himself.

One of Gerard's related habits of what could be called truculent citation is his tendency, not to name plants *after* women, but to characterize certain names as being only used by women:

> *L'Obelius* calleth it *Sulcatum* and *Gramen striatum*, or *Gramen pictum*: in English the Furrowed grasse, the White Chaemelon grasse, or straked grasse: and vsually of our English women it is called Ladies Laces, or Painted grasse. (25)

His incidental inclusion of names used uniquely by women incidentally confirms the special involvement of early modern women in plant culture. Aside from women, when citing botanical synonyms Gerard only tends to single out occupational groups (such as apothecaries) and national language groups (like that of "Italie"), or occasionally regional ones (like Cheshire).[7] Women are thus clearly conceived of as an equally distinct botanical community, speaking a separate language; yet such women are not recognized as "authors" but as anonymous contributors. Gerard refers almost always to "women" in a generic sense; while he names many male colleagues, he mentions only one woman by name in the entire herbal. He does describe an "experiment [that] was practiced by a worshipfull gentlewoman called mistresse *Anne Wylbraham*, vpon diuers of hir poore neighbours with good successe" (315). This example clarifies why women developed, as Gerard suggests, a botanical language all their own: they handled plants frequently in their work as local healers (Hoeniger, *Medicine* 17–18). But women's herbal work was under assault throughout the sixteenth century by their competitors in the College of Physicians (Hoeniger, *Medicine* 28). In conversations among educated herbalists, the adjective "ignorant" came increasingly to preface the word "women" or "herbwife." Gerard's general attitude to female herbalists may be summed up in his reference to a mistake made by most herbalists (presumably men), characterized in the ironic mode as an example of "such ignorance as is not amongst the simplest women" (319). Elsewhere Gerard refers to herbalists as "Symplist[s]" (95), those who study simples. The "simplest women" were also "Symplists" made responsible for the health of their communities without the privilege of much formal education. While Gerard excludes women's personal names from his herbal, he includes their botanical names, so that his book becomes an accidental anthology of early modern women's writing in the Book of Nature.

Gerard often shows women's botanical names to possess the very kind of metaphorical etymologies in which he takes a personal delight. He notes the soundness of the metaphor behind Snapdragon, the flowers of which are "fashioned like a ... dragons mouth; from whence the women haue taken the name" (438). He emphasizes the logic of another feminine name when he writes that "Swallow woort" has "long sharpe pointed cods, stuffed full of a most perfect white cotton resembling silke, as well in shewe as handling, which cods the grauer hath omitted in the figure: (our London gentlewomen haue named it Silken Cislike)" (750). He writes of a species whose leaves are "very much resembling the leafe of an Oke, which hath caused our English women to call it Okies of Ierusalem" (950). Elsewhere the naming practices of women are shown to have etymologies based in medical practice or other practical uses, rather than visual resemblance:

---

7 Indeed sometimes Gerard cites the regional botanical dialects of women, writing of one plant that "our London women do call it Rosearubie" (310) and of another that "women that dwell by the sea side, call it in English blew Daisies, or blew Camomill" (334).

> Shauegrasse is not without cause named *Asprella,* of his ruggednesse, which is not vnknowen to women, who scowre their pewter and wooden things of the kitchen therewith ... and thereupon some of our huswives do call it Pewterwoort. (958)

Gerard shows no objection to "huswives" using this name; in a later chapter, however, he expresses his conviction that a plant, and thus its name, is being misapplied by women in a medical context. He writes of a mushroom called great Toothwort or Clownes Lungwoort,

> There is nothing extant of the faculties heereof: Neither haue we any thing of our owne experience; onely our countrey women do call it Lungwoort, and do vse it against the cough, and all other imperfections of the lungs: but what benefit they reape thereby I know not; neither can any of iudgement giue me further instructions thereof. (1388)

Women (or a regional subgroup of them) are specifically excluded from being considered "of iudgement" about the medical applications of plants. Their name is not only subject to Gerard's skepticism (in terms of its relevance to the plant's medical virtues): it is transformed by being prefaced by him with a class marker ("Clownes") that denigrates both the name and the cure.

At the other end of the scale from these practical names given by women are those that appear the most abstract: Gerard speaks of a plant which "[w]e cal ... in English Pennie flower, or money flower, siluer plate, Pricksong woort, in Northfolk Sattin, & white Sattin, & among our women it is called Honestie" (378). The abstraction of the feminine name contrasts markedly with the visual or textural analogies embodied in most of the other names Gerard cites (barring "Pricksong woort," which may derive from a practical application). A similar sort of poetic abstraction by women may be found in a kind of spurge called "Cypress Spurge, or among women, Welcome to our house" (407); the plant is personified through its name, being given a voice. This may have been a common stylistic tactic among female plant-namers, since in another of Gerard's citations of feminine naming, the plant is similarly personified and given a voice: a plant which is called "in English spotted Sanicle" is also called "of our London dames prattling Parnell" (645). While Gerard does not pass adverse judgment on the second name, his phrasing might be read as calling into question whether "London dames" speak English. Women are seen to achieve similarly poetic effects in coming up with botanical names for a kind of "Cotton weed being of greater beauty then the rest":

> the flower being gathered when it is yong, may be kept in such manner as it was gathered, I meane in such freshnesse and well liking, by the space of a whole yeere after in your chest or elsewhere; wherefore our English women haue called it Liue long, or Liue for euer, which name doth aptly answer his effects. (517)

Here and elsewhere, the names that women give are often more poetically complex than those Gerard uses as their common English names; for instance, Gerard refers to "[t]his fruitfull or much bearing Marigolde, [which] is likewise called of the vulgar sort of women Iacke an apes a horse backe" (602); a little later, perhaps mistakenly, he refers to a different flower "well known by the name of Oxelip ... of which kinde, we haue one lately come into our gardens, whose flowers are curled and wrinckled after a most strange maner, which our women haue named, Iacke an apes on horsebacke" (635). The kind of articulated and simileic logic that often pervades Gerard's descriptions is embodied here in a single name.

That these names possess such poetic qualities may become less surprising in light of an anecdote related of the plant known as Setwall. Gerard writes that it

> hath beene had (and is to this day among the poore people of our northerne parts) in such veneration amongst them, that no brothes, pottage, or phisicall meates are worth any thing, if Setwall were not at one end: whereupon some woman poet or other hath made these verses:
>
> They that will haue their heale,
> Must put Setwall in their keale. (919)

Women are portrayed here as the typical authors of a certain kind of botanical verse. That it was quite conventional for women to compose such verse may be inferred from Gerard's casual attribution of it to "some woman poet or other," implying the poet's membership in a loosely understood community of like-minded herborizing and versifying women. A great deal of popular knowledge about healing and herbs had long been passed down in verse of this kind, so it seems significant that Gerard implies that female authors were characteristically responsible for its creation. If women were not only involved in growing, distilling, and prescribing herbs, but in poeticizing them too, that may explain why some of the first English women to put their names in print should have chosen for their textual collections such titles as *A Sweet Nosgay, or Pleasant Posye: Contayning a Hundred and Ten Phylosophicall Flowers* (Whitney) and *A handfull of holesome (though homelie) hearbs, gathered out of the goodlie garden of Godsmost holie word; for the common benefit and comfortable exercise of all such as are deuoutlie disposed* (Wheathill). Their foremothers had already made such a connection between *herbis* and *verbis* in their own work with words.

John Gerard was a surgeon, a gardener, and a herbalist—not a poet. His herbal, however, shows him placing an unexpected priority on pleasure and artistry as part of the value of herbalism. In his dedication to William Cecil, for instance, he asserts both the delightfulness and usefulness of plants—but he places delight first, and emphasizes that it is the artfulness of nature that is so delightful:

> if delight may prouoke mens labour, what greater delight is there than to behold the earth appareled with plants, as with a robe of imbroidered worke, set with orient pearles, and garnished with great diuersitie of rare and costly iewels? If this varietie and perfection of colours may affect the eie, it is such in herbes and flowers, that no *Apelles*, no *Zeuxis* euer could by any art expresse the like; if odours, or if taste may worke satisfaction, they are both so soueraigne in plants, and so comfortable, that no confection of the Apothecaries can equall their excellent vertue. But these delights are in the outward senses: the principall

> delight is in the minde, singularly enriched with the knowledge of these visible things, setting foorth to vs the inuisible wisedome and admirable workmanship of almightie God. (A2r)

Gerard here compares his sensory impressions of the green earth to those yielded by an embroidered cloth, by a painting truer than any *trompe-l'oeil* that Apelles and Zeuxis could produce, and by an artfully-prepared "confection of the Apothecaries." Moreover, he suggests that the principal pleasure of a plant, like that of a poem, is "in the minde." Admittedly, he concludes his paean by suggesting that "[t]he delight is great, but the vse greater, and ioyned often with necessitie"; but after only three sentences on the utility value of plants, he returns as if irresistibly to delight:

> And here beside the fruit, to speake againe in a word of delight; gardens, especially such as your Honor hath, furnished with many rare simples, do singularly delight, when in them a man doth behold a flourishing shew of sommer beauties in the middest of winters force, and a goodly spring of flowers, when abroad a leafe is not to be seene. (A2v)

In his address to the reader, too, Gerard elaborates with fervour on the sensory pleasures of plants:

> so necessarie and delectable, that nothing can be confected, either delicate for the taste, daintie for smell, pleasant for sight, wholesome for bodie, conseruatiue or restoratiue for health, but it borroweth the relish of an herbe, the sauour of a flower, the colour of a leafe, the iuice of a plant, or the decoction of a roote.

In concentrating on the delightful and sensuous value of plants and on their ability to produce summer's pleasures in the midst of winter, Gerard's herbal starts to resound with echoes of prefaces for collections of poetry that present themselves under the guise of a garden of verse. They, too, celebrate the combined delightfulness and utility of their contents;[8] they compare the vividness of their textual contents to painted pictures; they are of course artistic products; and they also trumpet the powers of a book to display within its pages the vivid pleasures of summer even on the darkest winter day. Appropriately, Gerard locates the proper habitat of the poet in the garden: "Whither did the Poets hunt for their sincere delights, but into the gardens of *Alcinous*, of *Adonis*, and the orchards of *Hesperides*? Where did they dreame that heauen should be, but in the pleasant garden of *Elysium*?"

Gerard elsewhere implies that flowers, like poems, can and should be transformative of those who experience them. In one chapter he moves from a description of violets to an expostulation on all beautiful flowers:

---

[8] Bushnell has also noted the prevalence of the pairing of "profit and delight" as "terms most familiar to us from the language of Renaissance poetics yet also omnipresent in garden manuals" (*Green* 93).

> the recreation of the minde which is taken hereby, cannot be but verie good and honest: for they admonish & stir vp a man to that which is comely & honest; for flowers through their beautie, varietie of colour, and exquisite forme, do bring to a liberall and gentlemanly minde, the remembrance of honestie, comelinesse, and all kindes of virtues. For it would be an vnseemely and filthie thing, as a certaine wise man saith for him, that doth look vpon and handle faire and beautifull things, and who frequenteth and is conuersant in faire and beautifull places, to haue his minde not faire, but filthie and deformed. (698–9)

Gerard's use of the word "places" here brings to mind its contemporary overlap of textual and territorial connotations—much like those of its frequent prefix, "common." Gerard here expresses a theory of the effect of floral beauty which may well have contributed to the popularity, in the contemporary poetry market, of the titular metaphor of the garden of verse: it would be to a bookseller's advantage if poetry could be construed as a "place," the beauty or even "honestie" of which would reflect upon whomever "look[ed] vpon" it.

The botanical aesthetic that Gerard describes also resembles the early modern collector's aesthetic. Indeed, as an adjunct to his persona as herbalist, Gerard adopted the persona of collector; he emphasizes in his dedication that "[t]o the large and singular furniture of this noble Iland, I haue added from forren places all the varietie of herbes and flowers that I might any way obtaine" ($A2^v$). Like most collectors of his day, Gerard exhibits a preference to describe in detail only the most rare and marvellous plants. He actively neglects descriptions of those that are very familiar; he writes, for instance, "The common garden Spurge is best knowne of all the rest, and most vsed; wherefore I will not spende time about his description" (405). We may infer, based on his logic in this and other examples, that those plants that he does describe in detail are more like the marvels, rarities and novelties favoured by collectors of both *naturalia* and *artificialia* (Daston and Park; Kenseth). Gerard singles some such marvels out for special remarks. In discussing what we would call cacti, he begins the description of "The Hedgehogge Thistle" by allowing himself to be flabbergasted: "Who can but maruell at the rare and singular workemanship which the Lord God almightie hath shewed in this Thistle[?]" (1013). Similarly, he declares of another that "[t]here is not amongst the strange and admirable plants of the worlde any one, that giueth more cause of maruell, or more mooueth the minde to honor and laud the Creator, then this plant" (1015). The reigning aesthetic that favoured the rare and marvellous affected the economics of herbalism, as can be seen by Gerard's account of the fluctuating value placed by the public on Golden Rod:

> It is extolled aboue all other herbes for the stopping of blood in sanguinolent vlcers and bleeding wounds; and hath in times past been had in greater estimation and regarde than in these daies: for within my remembrance, I haue knowne the drie herbe which came from beyond the seas, solde in Bucklers burie in London for halfe a crowne an ounce. But since it was founde in Hampsteed wood, euen as it were at our townes end, no man will giue halfe a crowne for an hundred weight of it: which plainly setteth foorth our inconstancie and sudden

> mutability, esteeming no longer of anything (how pretious soueuer it be) than whilest it is strange and rare. (349)

A decade earlier, William Harrison had lamented the same phenomenon in his *Description of England* when he discussed plants "which are most common by reason of their plenty, and most vile because of their abundance" (268):

> we have neglected our own good gifts of God, growing here at home, as vile and of no value, and had every trifle and toy in admiration that is brought hither from far countries, ascribing I wot not what great forces and solemn estimation unto them, until they also have waxen old, after which they have been so little regarded, if not more despised, amongst us than our own. (263)

Unexpectedly, even the stock market in herb sales at Bucklersbury embodied the aesthetic of rarity that pervaded European collecting. Herbalists themselves were susceptible, it seems, to the tendency to prefer the exotic over the domestic cure, and there was a special vogue for medicines from newly discovered plants (Wear, "The Early Modern Debate"). Gerard alludes to an addiction to novelty among herbalists in his description of a plant he calls "One sommers Nauell woort":

> which plant was giuen to *Mathiolus* by *Cortusus*, who (as he affirmeth) receiued it from Syria; but I thinke he saide so to make *Mathiolus* more ioifull: but surely I surmise he picked it out of one olde wall or other, where it doth growe euen as the small Chickweede, or Nailewoort of the wall do. (425)

In the next section, on the ecological niche of the same plant, Gerard makes another jab at Mattioli when he notes that it "groweth vpon olde stone and mudde wals: notwithstanding I haue (the more to grace *Mathiolus* great iewell) planted it in my garden" (425). Gerard may mock, but he too was a collector of what he hoped were genuine exotics, as is evinced by the number of times he describes a plant as "a stranger in England," only to conclude with audible pride, "it groweth in my garden" (628). One consequence of this widespread botanical aesthetic would be that Gerard's technically utilitarian physic garden may not have appeared much different from those created by amateurs interested in plants for purely aesthetic purposes.

Yet marvellousness and rarity were not the only fashionable aesthetic qualities which a herbal like Gerard's might embody. Equally important as an artistic principle, particularly for the collector, was variety (Kenseth 44). The plants themselves were remarkably varied, with species turning up in all shapes and sizes, serving all sorts of functions, growing in all sorts of ways. Gerard himself describes one virtuouso gardener,

> curious in Graffing and Planting of Fruites, who hath in one peece of ground, at the point of three score sundrie sorts of Peares, and those exceeding good, not doubting but if his minde had beene to seeke after multitudes, he might haue gotten together the like number of those of worse kindes. (1267)

At a time when there "yeerely commeth to our handes others not before known" (1311), there appeared to be no limit to the possibilities of botanical variety, which formed a large part of the appeal of plants to collectors in the period. Conveniently, the inevitable nature of an encyclopedic herbal was to be as varied as possible, each page promising a new plant, such that the very bibliographic format tended to highlight the botanical variety contained therein. It was not simply that herbals contained so many different plants, but that the ordering principles behind most sixteenth-century herbals tended to emphasize their differences. This was not intentional but an inadvertent consequence of the alphabetization that most commonly formed the ordering principle. Gerard diverges from this mode and takes up a series of categories that follow the physical cues offered by plants: he begins with grasses and rushes and proceeds through similar categories in as orderly a way as possible. But the order is not, after all, very orderly—particularly toward the end, when he clumps together all the plants he has not yet discussed, including many of relatively recent arrival in Europe. The very lack of clear categories for the plants he discusses can make for the most incongruous sequences. At one point he moves from looking at cacti—then exotic, barely known or namable—to a consideration of clover, than which few weeds could be more common. He turns from pages containing plants very little known in Europe to the familiar "Sun deaw" and the "Fen berries" known to grow "especially in Cheshire, and Staffordshire, where I haue found [them] in great plentie" (1364–7). A little later is a sudden shift from English mushrooms to trees that grow in Timor and Java (1388–9). The resulting layout, from page to page, breaks all natural ecological rules; only a paper-garden could accommodate such an eclectic arrangement. Such sequences embody, moreover, the desirable aesthetic quality of incongruous variety—and make his herbal resemble, in this respect, the poetic anthologies that aimed for a similar variety in the arrangement of their contents.

There are basic similarities between the verbal tasks undertaken by poets and herbalists and the rhetorical techniques by which they completed them. Like a poet, Gerard strives in his prose descriptions to bring the impression of a material thing or sensory experience into the minds of readers who might never have encountered it. In the case of the poet, this approach is taken at least partly because the thing written of is often fictive, and so needs to be likened to familiar material things in order to be understood. In Gerard's case, the materials being described are not (usually) fictive, but simply unusual or unfamiliar to his anticipated audience. Both poet and herbalist, in response to this task, avail themselves of figurative language. Both, moreover, commonly conceived of their texts in relation to the faculty of memory. Mnemosyne was the mother of the muses, while for herbalists the ability to remember the distinctions among different plants was a critical skill. In poetry composed for oral performance, rhetorical figures and formulae could function as cues to jog the speaker's or singer's memory; as has been shown, a parallel tradition of oral herbal verse was embedded in Gerard's text. Although Gerard's text was a reference book, it was too large to be easily carried about on field expeditions, so it would best fulfill its function if its descriptions were so vivid as to make the plants they describe unmistakably and distinctly memorable to a reader. Methods for making memorable descriptions were themselves described with great specificity in rhetorical manuals of the period; in the standard schoolroom *Rhetorica Ad Herennium*, for instance, it was written that

> We ought ... to set up images of a kind that can adhere longest in memory. And we shall do so if we establish likenesses as striking as possible; if we set up images that are not many or vague, but doing something; if we assign to them exceptional beauty or singular ugliness; if we dress some of them with crowns or purple cloaks, for example, so that the likeness may be more distinct to us; or if we somehow disfigure them, as by introducing one stained with blood or soiled with mud or smeared with red paint, so that its form is more striking, or by assigning certain comic effects to our images, for that, too, will ensure our remembering them more readily. (Cicero III, xxii, 37)

As shall be seen, all of these tactics are undertaken by Gerard in the course of his descriptions of plants.

The genre of herbal writing did not only partake of poetical resources; the general shape of its content also bore an unusual resemblance to the rhetorical unit most widely prescribed in the humanist educational regimen, the commonplace, a type of text "characterized by brevity, prescriptiveness, and strong closure" (Crane 8). Those three descriptors could equally characterize Gerard's rhetoric of description as he inherited it from the herbal tradition. The ideal of brevity is embodied by the short sub-sections into which he divides each chapter, breaking up what could be a long discourse into bite-sized pieces; his work in the sections detailing the "vertues" of each plant is even set in a kind of point form, with each separate curative power alphabetically bulleted in the margins. The declarative nature of Gerard's descriptions are also well characterized by what Crane calls "strong closure"—sometimes to the point of arbitrarily cutting off what Gerard dismisses in other herbal texts as mere "cauilling" and "controuersie" (95, 149, 840). The third distinctive quality of the commonplace—prescriptiveness—is of course at the heart of medicinal herbalism.

Gerard thus both echoes the stylistics of the commonplace book and also demonstrates a kind of herbal poetics of his own in his vivid appeals to the senses and his reliance on figurative language in order to evoke the particular presence of an absent or unfamiliar thing, whether real or imagined. Gerard's very first entry demonstrates the precision and vivacity of his descriptive rhetoric:

> Common Medow grasse hath very small tufts of rootes, with thicke hairie threds depending vpon the highest turfe, matting and creeping on the ground with a most thicke and apparent shew of wheaten leaues, lifting vp long, thin, iointed and light stalks, a foote or a cubite high, growing small and sharpe at the top, with an eare which is loose and hanging downward, like the tuft or top of the common Reed called *Vallatoria*. (1)

As in the anatomizing *blazons* of the sonneteers of the 1590s, Gerard isolates the concrete details of the plant, even without the assistance of a technical botanical vocabulary; he separately depicts the roots, leaves, stalks, and ears in this single brief description, rather than presenting an anatomically-cohesive image. He starts at the roots and works his way up to the top in an orderly fashion that mimics the directional sweep of the eye upon first encountering an uprooted specimen such as one might encounter in a herbarium. He uses ordinary but apt adjectives, singly and in conjunction, to modify almost every noun. He also musters a variety of verbs in a tense that conveys a sense of liveliness and even agency in what might otherwise appear a static picture: the grass is portrayed as if caught in the act of "creeping on the ground" and "lifting vp long ... stalks." The agency Gerard assigns the plants through these verbs contributes not only to the vividness of his description—the images are, as recommended in *Ad Herennium*, "doing something"—but also to their personification.[9]

Beyond the ordinary vividness that comes from an exacting description of concrete detail, enlivened by a judicious use of adjectives and verbs, Gerard's method of word-painting often extends to still more poetic approaches. I have mentioned his tendency to personify plants; more pervasive, however, is his rhetoric of similitude. An example may be found in his account of "Cats taile," which "hath long, rough, thicke and flaggie leaues, full of a spungious matter, like marrow, among which leaues, growth vp a long smooth naked stalke, without knot or ioint, fashioned like a speare ..." (41–2). In the first case Gerard's simile serves to describe the texture of an unusual material,[10] while in the second the simile describes its shape. In both cases, Gerard turns to figurative language in order to increase the sensory impact of his words.[11] The turn to similitude was a

[9] Other examples of these techniques abound. The vivid adjectival phrases in the "crooked and crambling rootes" of the Cypresse grasse (19) and the "thrummie threds" of the root of Spike Flote grasse (13) also turn alliteration to good effect. In another instance of the liveliness with which his verbs imbue his specimens, he describes the root of Flote grasse, "from the which riseth vp long and crooked stalks, crossing, winding and folding one within another .... At the top of these stalks and somewhat lower, there do thrust foorth many tufts or tassels, of an ouerworne reddish colour" (13). Elsewhere, of the flowers of Water Hempe, he writes "The flowers ... stand perting vpon the top of the sprigs, which at length vanish away into downe ..." (574). He also shows a marked penchant for personification in his description "Of the Sorrowfull tree, or Indian Mourner," whose flowers bloom only at night, and "in the day time looke withered and with a mourning cheere: the leaues also at that time shrinke in themselues together ... very sadly lumping, lowring, and hanging downe the head, as though it lothed the light, and could not abide the heate of the sunne" (1343).

[10] Another vivid example of a textural simile may be found in his description of sage leaves, which he depicts as "like in roughnes to woolen cloth thread bare" (622). That Gerard does not stop with the analogy to wool, but makes the further specification of its "thread bare" state, shows the care with which he formed his analogies.

[11] There are many other vivid examples of simile in Gerard's descriptions of a plant's physical appearance. Speaking of a "strange kinde of Woodbinde," he notes that the flowers are "somewhat dasht with yellow, by little and little stretched out like the nose of an Elephant" (744). Of another plant he says that "[t]he three leaues that hang downward are like a gaping hood" (51), while "[t]he yellow Lillie hath very long flaggie leaues, chamfered

rhetorical choice that Gerard and his contemporaries in herbalism were notable for making (Arber 160). Arber suggests that "[t]his amateur method was the only choice so long as the possibility of measurement was disregarded, and there was no terminology by means of which shape and arrangement could be expressed by reference to an agreed botanical standard" (162). Yet as Arber herself suggests, measurement was not unavailable but "disregarded" by Gerard and many of his contemporaries; and the unavailability of standard terms could have been righted with relative ease through the kind of neologism that Turner, at least, was not reluctant to practice when it came to plant names. But Gerard, while aware of an extant Latin vocabulary for "the differing names of their seuerall parts [the parts of plants]," concludes that such terms are "more ... than our vulgar toong can well expresse" (1), and on these grounds chooses not to use them at all. He shows no such reluctance in his flourishing use of figurative language.

Such a rhetorical choice would not remain uncontroversial for long in natural history; shortly after Gerard's time, calls came to eliminate just such figurative logic from knowledge-making disciplines. Until then, though—and until the accompanying arrival of a clearer separation of art and science—similitude was recognized as one of the most powerfully evocative or persuasive figures available to either a prose writer or a poet, as Puttenham makes clear in his *Arte of English Poesie*:

> As well to a good maker and Poet as to an excellent perswader in prose, the figure of *Similitude* is very necessary, by which we not onely bewtifie our tale, but also very much inforce and inlarge it.
>
> I say inforce because no one thing more preuaileth with all ordinary iudgements than perswasion by *similitude*. (247–8)

Puttenham places this superlative description of similitude as a poetical figure immediately after his description of description itself as another poetical figure. That description itself was conceived of at the time *as* a trope shows the extent to which Gerard's task was not merely a matter of common sense. The need to turn to

or channeled, hollow in the middest like a gutter" (91). Hyacinth has leaves "hollow like a little trough" and petals "spread abrode like a star" (97). Elsewhere: "The flowers grow at the top, thrust or packt together like a bunch of Grapes, of a pleasant bright skie colour, euerie little bottle-like flower, set about the hollow entrance with small white spots, not easie to be perceiued" (104). Speaking of Grecian Mustard, he describes the "round flat huskes or seede vessels, set vpon the stalke by couples, as it were sundry paires of spectacles" (206). Still elsewhere: "euery leafe doth resemble in colours the most faire and beautifull feather of a Parrat"; later in the same chapter, "I can compare the shape thereof [of the flower] to nothing so fitly as to the veluet head of a Stagge, compact of such soft matter as is the same" (225). He describes the Barren woort's leaves as "somewhat snipt about the edges, and turning themselues flat vpright, as a mann turneth his hande vpwardes when he receiueth money" (389). The flowers of Medick fodder "turne into round wrinckled knobs, like the water snaile, or the fish called Periwinck, wherein is contained the seed, fashioned like a little kidney" (1029).

similitude to support the success of "Hypotiposis, or the counterfeit representation" becomes clear after reading Puttenham's description of description:

> The matter and occasion leadeth vs many times to describe and set foorth many things, in such sort as it should appeare they were truly before our eyes though they were not present, which to do it requireth cunning: for nothing can be kindly counterfait or represented in his absence but by great discretion in the doer. (245)

By emphasizing and even mystifying the describer's ability to raise up the images of things "before our eyes though they were not present," Puttenham defamiliarizes the very ordinary act of description and reminds us of its artful nature. He goes on to contrast the description of things "not naturall or not veritable" with that of "true" things, and finds the former act, not surprisingly, the more difficult; nonetheless, Puttenham also asserts that "nothing can be kindly counterfeit"—including "true" things, such as the plants Gerard describes—without "great discretion." One of Gerard's primary rhetorical tasks, then, was far from merely commonsensical in the period; instead, description itself was figured as a kind of poetical ornament requiring the cunning deployment of other ornaments, such as the similitude, which Puttenham goes on to define.

While it might be hard to see what "cunning" lies in the previous examples of Gerard's similes, his figurative logic at times becomes quite complex, exceeding the simple formula of "A is like B." Sometimes a single simile is articulated to provide a parallel image of more than one anatomical part of the plant, so that the simile tends to function as a sort of allegorical system, as in his description of "Herbe Iue": "The flower gapeth like the mouth of a beast, and hath as it were a white toong, the lower and vpper iawes are white likewise, spotted with many bloudie spots ..." (422); as if persuaded by his own animal imagery, the simile shifts into more direct metaphor in the final phrase with the appearance of blood. Likewise, in speaking of a kind of Fumitorie, Gerard exceeds simile and turns to a metaphorical identification in this segmented conceit: "the flowers hereof are as it were small birdes, the bellies or lower parts wherof, are of a white colour ..." (932). In these two examples, his explanation of the anatomy of plants appears to assume a prior and better understanding of the anatomy of animals. Similarly, in a valiantly redoubled attempt to describe the flowers on the plant known to him as White Dogs stones, his simile becomes a matter of articulated correspondences which themselves are finally shown to correspond as a whole to a conventional artistic representation:

> The small flowers are like an open hood or helmet, hauing hanging out of euerie one as it were the bodie of a little man without a head, with armes stretched out, and thighs straddling abrod, after the same maner almost, that the little boies are woont to be pictured hanging out of Saturnes mouth. (157)

Gerard's references thus take in not only physical concordances but allusions to iconic works of art, showing the range of imagination that he brings to his work.

Moreover, the "singular ugliness" of the image corresponds well to one of the descriptive strategies recommended in the *Ad Herennium*.

Gerard's imagination outdoes itself in a different way when he describes the scent of the Day Lilly in the evening, which he finds to be "as rotten and stinking, as if it had been trodden in a dunghill a moneth together in foule and rainy weather" (91); presumably this likeness is based on an imaginative exaggeration of actual experience, rather than any concrete sense of the specified smell. A similarly complex mingling of concrete specificity and wild imaginative work may be found in the similes under the chapter heading for "Calues Snout, or Snapdragon"; while he labels all the figures of the varieties of this plant as types of Snapdragon, and acknowledges a couple of other possible metaphorical names, he expresses a strong preference for a strikingly idiosyncratic metaphor of his own envisioning:

> The flowers growe at the top of the stalks of a purple colour, fashioned like a frogs mouth, or rather a dragons mouth; from whence the women haue taken the name Snapdragon. The seede is blacke, conteined in long husks fashioned like a calues snout, whereupon some haue called it Calues snout, or in mine opinion it is more like vnto the bones of a sheepes head that hath beene long in the water, the flesh consumed cleane awaie. (438)

As in the last example, the final part of his description here might amount to an example the technique recommended in the *Ad Herennium* that is said to make an image more memorable "if we somehow disfigure" it "so that its form is more striking." Similarly, in his attempt to evoke the appearance of the Common Reede, he creates two sorts of similes, the first very straightforward—comparing one plant to another—and the second entirely the product of his own imagination:

> The Common Reede hath long strawie stalks, full of knottie ioints or knees like vnto corne, whereupon do grow very long rough flaggie leaues. The tuft or spokie eare doth grow at the top of the stalks, browne of colour, barren and without seede. And doth resemble a bush of feathers ... (32)

Since it is unlikely that even Gerard—famous for his eye-witness account of the "tree bearing Geese" (1391)—had ever actually seen a "bush of feathers," this compound metaphor indicates the lengths to which he would stretch his imagination in poetic directions in order to accomplish the most basic descriptive task.

Sometimes, despite his best poetic efforts, Gerard is not satisfied with the likeness he produces. For instance, he describes the leaves of the Swolen Colewoorte as "bunched or swollen vp about the edges as it were a peece of leather wet and broiled on a grid iron, in such strange sort that I cannot with words describe it to the ful" (247); Gerard feels that words fail despite the vividness of his entirely imaginary comparison. Another simile shows the primary shortcoming of the trope as a method of description when he writes of a kind of iris that "vnder ech of these leaues appeere three brown aglets, like the toong of a small bird" (52); in this case, although Gerard appears quite pleased with it, his chosen vehicle is surely at least as opaque as his tenor in its presumption of a remarkably specialized knowledge of fauna. He leaves traces in his text of the difficulty of his task of translating a visual image into words: "I cannot with words set foorth, such is the sundrie mixtures of colours that nature hath bestowed in hir greatest iollitie vpon this flower" (255). He more than once goes so far as to portray as a kind of high art the task of capturing in words something that, according to him, is barely susceptible to that medium; this is especially true of plants he admires greatly. At one point, he almost sounds like a conventional poet-lover confounded by the ineffable beauty of his beloved, which he personifies (both in a name straight out of a pastoral or romance and in his pronoun reference) here:

> It farre exceedeth my skil to describe the beautie and excellencie of this rare plant called *Floramore*; and I thinke the pencil of the most curious painter will be at a staie when he shall come to set him downe in his liuely colours: but to colour it after my best maner this I saie ... (255)

His description, in this instance, correlates to the suggestion in the *Ad Herennium* to "assign to [the described thing] exceptional beauty." Here, Gerard connotes the task of verbal description as a kind of painting. Elsewhere, too, he surrenders his verbal task in favour of a visual one:

> to describe the colour in words, it is not possible, but this waie; laie vpon paper with a pensill a yellowe colour called Masticke, which being drie, laie the same ouer with a little saffron steeped in water or wine, which setteth foorth most liuely the colour. (610)

Even his sense of the limitations of the power of words shows his sensitivity to his task as a translator of the visual into the verbal.

In speaking of a plant called Ladies Laces, Gerard shows his conviction that such simileic logic as he himself favours is at the base of much herbal etymology: the plant is so named, he tells us quite reasonably, "because it is stript or furrowed with white and greene strakes, like silke laces" (5). Elsewhere, if it is ambiguous whether the etymology of an English plant name is based on a utilitarian logic—describing the plant's "vertue"—or on a metaphorical logic, Gerard tends to insist that the latter is what matters: a plant is called "Fleawoort, not because it killeth fleas, but because the seeds are like fleas" (471). He solves (at least, to his own satisfaction) one botanical dispute by a similar means when writing of the sunflower:

> The flower of the Sunne is called in Latine *Flos Solis*, taking that name from those that haue reported it to turne with the sunne, the which I coulde neuer obserue, although I haue endeuored to finde out the truth of it; but I rather thinke it was so called because it doth resemble the radiant beames of the Sunne ... (614)

Gerard also evinces the poetic orientation of the sixteenth-century herbalist's rhetorical work when he demonstrates the same tendency as William Turner did to care less about the exact wording of a name than about the metaphor at its heart. To select just one example among many, he speaks of a kind of moss "set forth vnder the title of *Muscus Pyxidatos*, which I haue englished Cup Mosse, or Chalice Mosse" (1371). Here we see poetic logic built into his botanical subject matter at its most fundamental and (for sixteenth-century botanists) preoccupying level: that of naming.

Imaginary Gardens: Poets as Herbalists

Gerard often shows a marked proclivity to turn to writers better known for their poetic output than for their botanical know-how. This is yet another example, not only of the anthological nature of Gerard's herbal, but of its tendency to bridge the gap between poetry and plants. His inclusion, for instance, of the poem about Setwall by "some woman poet or other" (919) is not his only acknowledgment of the long-standing tradition of intermingling plants and poetry in verse that was valued primarily for its mnemonic function (Johnston, *Cleveland's* 4). In his discussion of Linaria, he writes that "the whole plant so much resembleth *Esula minor*, that the one is hardly knowne from the other, but by this olde verse: *Esula lacte scit, sine lacte Linaria crescit*" (440). A similar example functions in both Latin and English through the retention of its rhyme after translation:

> *Pliny* calleth it *Euphrosmum*, because it maketh a man merrie and ioyfull: which thing also the olde verse concerning Borage doth testifie:
> Ego Borago gaudia simper ago.
> In English.
> I Borage bring always courage. (654)

Likewise, Gerard writes that "[t]he powder of the seede of Fennell, drunke for certaine daies together fasting, preserueth the eyesight, whereof was written this distichon following,

> *Foeniculus, Rosa, verbena, Chelidonia, Ruta,*
> *Ex his fit aqua quae lumina reddit acuta.*
> Of Fennel, Roses, Veruerine, Rue and Celandine,
> Is made a water good to cleere the sight of eyne. (877)

In all of these cases, the plant is understood through the medium of poetry; moreover, the poetry is thought able to "testifie" the truth about the plant. While these are anonymous poems, Gerard elsewhere assigns this sort of authority to named poets, including those who were not primarily writing about plants.

One section in which "the Poets," anonymous and otherwise, tend to turn up quite frequently is that in which Gerard assembles all the known names of the featured plant: "Rocket is called ... in Latine *Eruca*: in high Dutch *Rauekenkraut*: in French *Roquette*: in lowe Dutch *Rakette*: in Italian *Ruchetta*: in Spanish *Oruga*: in English Rocket, and Racket. The Poets do oftentimes name it *Herba Salax* ..." (192). Sometimes he goes further, not only including the names assigned by poets but explaining their context in a particular poem:

> Of this Cats taile *Aristophanes* maketh mention in his Dialog of Frogs, where he bringeth them foorth one talking with another, being very glad that they had spent the whole day in skipping & leaping *inter Cyperum* and *Phleum*, among Galingale and Cats taile. *Ouid* seemeth to name this plant *Scirpus*, for he termeth the mats made of leaues, Cats taile mats, as in his sixt booke *Fastorum*,
>
> At Dominus, discedite, ait, plaustroq morantes
> *Sustulit, in plaustro scirpea matta fuit.* (42)

When quoting poetry, Gerard follows the period convention of setting off the verse from the rest of the text, a layout that highlights the presence of the poetry such that, when any quantity is quoted, the page can look as if it belongs to a book of poetry rather than a herbal (Figure 4.1).

Gerard seems to set the authority of herbalists and that of poets on par when he presents their words contiguously, showing no preference for the account of one or the other. For instance, he uses the word of a poet to confirm the judgment of physicians in reference to the Latin name of garden Succorie: "in shops it is called *Cicorea*, which name is not onely allowed of the later Phisitions, but also of the Poet *Horace* in the 31. Ode of his first booke, *Me pascunt oliuae, / Me Cicorea, leuesque maluae*" (221). Elsewhere, to establish the habitat of the Rush Daffodil, he turns from the word of a herbalist to that of a poet, whose opinion wins the lion's share of textual space:

> *Dioscorides* saith that they be especiallie found vpon mountains. *Theocritus* affirmeth the Daffodils to growe in medowes in his 20. *Eidyl,* where he writeth that the goddesse *Europa* entering with hir Nymphs into the medowes, did gather the sweet smelling Dafodils in these verses.
>
> *But when the girles were come into*
> *The medowes flowring all in sight,*
> *That wench with these, this wench with those*
> *Trim flowers, themselues did all delight:*
> *She with the* Narcisse *goodin sent,*
> *And she with* Hyacinths *content.* (113)

The same pattern occurs in another chapter:

> *Dioscorides* writeth that all the Dockes being boiled doe mollifie the bellie, which thing also *Horace* hath noted in his second booke of *Sermons*, the fourth Satyre writing thus,
>
> *Si dura morabitur aluus*
> *Mugilus, & viles pellent obstantiaconchae,*
> *Et lapathi breuis herba.* (314)

Elsewhere he turns to Virgil:

> *Virgill* in his seuenth ecloge of his Bucolicks, maketh mention of *Baccharis*, and doth not onely shew that it is a garland plant, but also such a one as preuaileth against inchantments, saying,

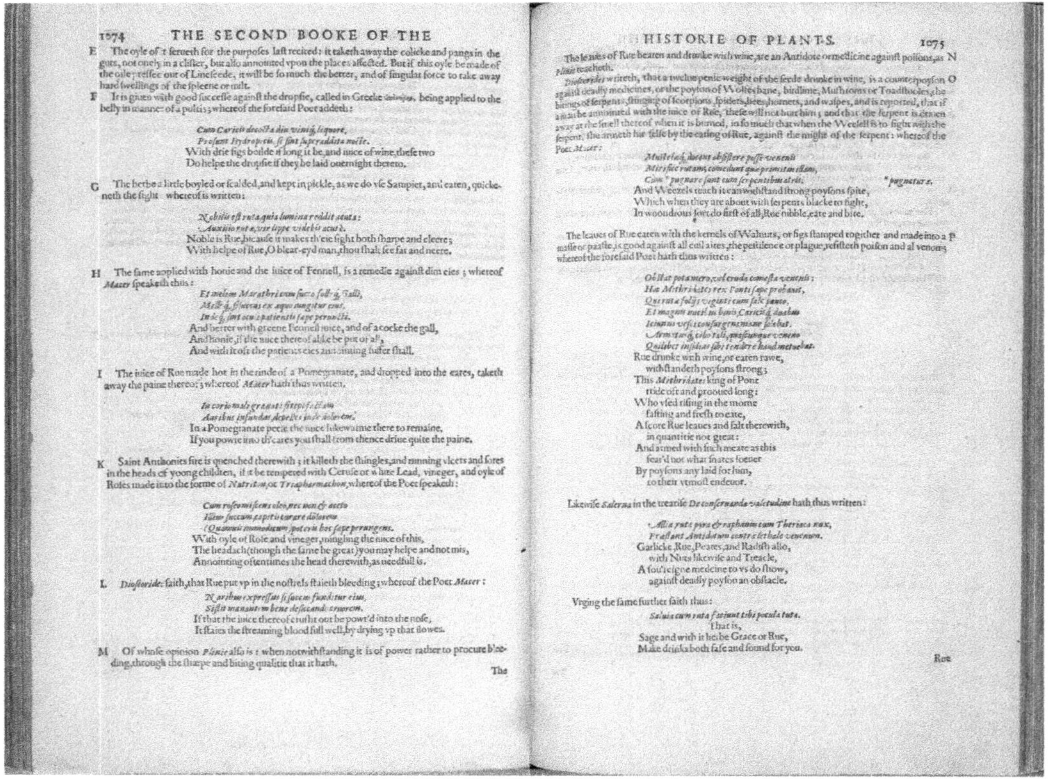

1074 THE SECOND BOOKE OF THE

E The oyle of it serueth for the purposes last recited: it taketh away the colicke and pangs in the guts, not onely in a clister, but also annointed vpon the places affected. But if this oyle be made of the oile pressed out of Lineseede, it will be so much the better, and of singular force to take away hard swellings of the spleene or milt.

F It is giuen with good successe against the dropsie, called in Greeke [illegible], being applied to the belly in manner of a pultis; whereof the foresaid Poet addeth:

With drie figs boilde if long it be, and iuice of wine, these two
Do helpe the dropsie if they be laid ouernight thereto.

G The herbe a little boyled or scalded, and kept in pickle, as we do vse Sampier, and eaten, quickeneth the sight: whereof is written:

Noble is Rue, bicause it makes the'ie sight both sharpe and cleere;
With helpe of Rue, O bleare-eyd man, thou shalt see far and neere.

H The same applied with honie and the iuice of Fennell, is a remedie against dim eies; whereof *Macer* speaketh thus:

And better with greene Fennell iuice, and of a cocke the gall,
And honie, if the iuice thereof alike be put of all,
And with it oft the patients eies annointing suffer shall.

I The iuice of Rue made hot in the rinde of a Pomegranate, and dropped into the eares, taketh away the paine thereof; whereof *Macer* hath thus written.

In a Pomegranate peele the iuice likewise there to remaine,
If you powre into th'eares you shall from thence driue quite the paine.

K Saint Anthonies fire is quenched therewith; it killeth the shingles, and running vlcers and sores in the heads of yoong children, if it be tempered with Ceruse or white Lead, vineger, and oyle of Roses made into the forme of *Nutritum*, or *Triapharmacon*, whereof the Poet speaketh:

With oyle of Rose and vineger, minging the iuice of this,
The headach (though the same be great) you may helpe and not mis,
Annointing oftentimes the head therewith, as needfull is.

L *Dioscorides* faith, that Rue put vp in the nostrels staieth bleeding; whereof the Poet *Macer*:

If that the iuice thereof crusht out be powr'd into the nose,
It staies the streaming blood full well, by drying vp that flowes.

M Of whose opinion *Plinie* also is: when notwithstanding it is of power rather to procure bleeding, through the sharpe and biting qualitie that it hath.

The

HISTORIE OF PLANTS. 1075

N The leaues of Rue beaten and drunke with wine, are an Antidote or medicine against poisons, as *Plinie* teacheth.

O *Dioscorides* writeth, that a twelue penie weight of the seede drunke in wine, is a counterpoyson against deadly medicines, or the poyson of Wolfes bane, birdlime, Mushroms or Toadstooles, the bitings of serpents, stinging of scorpions, spiders, bees, hornets, and waspes, and is reported, that if a man be annointed with the iuice of Rue, these will not hurt him; and that the serpent is driuen away at the smell thereof when it is burned, insomuch that when the Weesell is to fight with the serpent, she armeth her selfe by the eating of Rue, against the might of the serpent: whereof the Poet *Macer*:

And Weezels teach it can withstand strong poysons spite,
Which when they are about with serpents blacke to fight,
In woondrous sort do first of all, Rue nibble, eate and bite.

P The leaues of Rue eaten with the kernels of Walnuts, or figs stamped togither and made into a masse or pastle, is good against all euil aires, the pestilence or plague, resisteth poison and al venoms whereof the foresaid Poet hath thus written:

Rue drunke with wine, or eaten rawe,
withstandeth poysons strong;
This *Methridates* king of Pont
tride oft and prooued long:
Who vsed rising in the morne
fasting and fresh to eate,
A score Rue leaues and salt therewith,
in quantitie not great:
And armed with such meate as this
fear'd not what snares soeuer
By poysons any laid for him,
to their vtmost endeuor.

Likewise *Salerna* in the treatise *De conseruanda valetudine* hath thus written:

Garlicke, Rue, Peares, and Radish also,
with Nuts likewise and Treacle,
A souereigne medicine to vs do show,
against deadly poyson an obstacle.

Vrging the same further saith thus:

That is,
Sage and with it herbe Grace or Rue,
Make drinks both safe and sound for you.

Rue

Fig. 4.1 A two-page spread from John Gerard's *The Herball or Generall Historie of Plantes*

> *—Bacchare frontem*
> *Cingite, ne vati noceat mala lingua futuro.*
> Decke the forehead of the Prophet with plowmans Spiknard,
> To keepe him from an ill toong that hurtes without regard. (648)

What is especially interesting about these last three examples is that Gerard turns to poets not simply to confirm a plant's name—a verbal matter, about which poets might be expected to wield some authority—but to confirm matters related to their habitat or medicinal virtues, domains in which one would think herbalists might be the better sources. He is even able to cite an ancient herbalist citing an ancient poet: "as *Plinie* writeth, *Aristophanes* of olde, being an ancient comicall poët, witnesseth, that ointments were woont to be made of the roote thereof ..." (648). Gerard demonstrates that he came honestly by his habit of quoting poetry as botanical evidence when he shows that ancient herbalists did the same.

Some of the poets that Gerard quotes have better qualifications than others to comment on herbal matters with authority. It is perhaps reasonable to value equally the words of an author of an ancient natural history and the author of the original georgic: "*Pliny* in his 33.booke cap.5 maketh mention by the waie of this herb, and calleth it *Lute* ... and likewise *Virgil* in his Bucolickes, the fourth Egloge" (398). He similarly turns from Galen to the Virgil of the georgics:

> It is of a meane temperature in cooling and drying. *Galen* saith it doth moderately waste and consume, especially while it is yet soft and new gathered.
>
> That with the blew flower or purple, is thought to be that, which is of *Virgill* called *flos Amellus*: of which he maketh mention in the fourth booke of his Georgickes. (393–4)

But Gerard is not especially discriminating about citing only poets known for their contributions to botany. He cites the comedies of Plautus (667) and of Aristophanes (42, 648) and the love poetry of Catullus (540, 854–5). He cites Ovid and Homer (734), Hesiod and Horace (786). A single chapter turns into a veritable anthology of ancient poets—Ovid, Virgil, Catullus, Prudentius—all mustered in order to prove a point about a pine tree (1177–8).

Gerard sometimes sets the authority of poets even higher than that of herbalists, using the verse of the former to solve long-standing disputes among the latter. For instance, he speaks of the confusion about a plant's identity among ancient physicians and herbalists—"Some haue made a doubt whether *Maiorana* and *Sampsychum* be all one"—and explains the origins of the disagreement among Galen, Dioscorides, and Pliny (539), respectively the chief ancient authorities on medicine, herbalism, and natural history. He proceeds to cite two poets, Virgil and Catullus, demonstrating how their poetry can help to resolve the dispute:

> *Virgil* in the first of his *Aeneidos* sheweth, that the shrub *Amaracus* beareth flowers, writing thus,
>
> Ubi mollis Amaracus illum
> Floribus, & dulci aspirans complectitur umbra.

> Where beds of Time, and Marierome so soft,
> And lustie flowers in greene wood shade, him breathes in comforts oft.
> Likewise *Catullus* in Epithalamio, or marriage song of *Iulia* and *Mallius* saith,
> *Cinge tempora floribus*
> *Suaue olentis Amaraci.*
> Compas the temples of the head with flowers,
> Of Amarac affording sweete sauours. (540)

A little later, Virgil once again saves posterity from the errors of Dioscorides:

> *Dioscorides* writeth against all truth, that it neither beareth flowers nor seede: after *Theophrastus*, *Virgil* witnesseth that it doth beare flowers, in the 12. of his *AEneidos*:
> *Dictamnum genitrix Creataea Carpit ab Ida,*
> *Puberibus caulum folijs, & flore comantem*
> *Purpureo.*
> In *Virgill.*
> But *Venus* much appald at this hir sonnes vnwoorthie paine,
> Greene Dittanie from Ida sacred mount in Creta brings,
> The stalke with tender leaues, and blossome purple fresh that springs. (651–2)

Although Virgil is supposed to be depending upon another herbalist (Theophrastus), it is Virgil's account that is quoted and Virgil who gets the credit from Gerard. In both instances, Gerard shows poets (in his view) to be better botanists than botanists themselves.

Yet at other times Gerard shows an ambivalence about the connection between poetry and plants. This ambivalence sometimes verges on outright scorn for the very poetical etymology and evidence that, at other times, he includes with a perfectly straight face. In his chapter on Saffron, he suggests that

> [s]ome that are disposed to dissemble and iest with their friends and to make them merrie with pretie Poeticall figments, haue giuen it the name of a Damsell, whereof *Ouid* maketh mention, which to recite were impertinent to our historie. (124)

Only the chapter before, however, Gerard was happily quoting Martial's epigrams in an attempt to establish whether the name *Frittillaria* originally referred to a table at which chess was played or whether the game might have been dice—a debate that is surely the epitome of impertinence (123). His ambivalence is even more palpable when he dismisses the legendary magical powers of Moly:

> As for repeating of foolish and vaine figments, the coniuring of witches, & magicians inchantments, which haue beene attributed vnto those herbes, I leaue them to such as had rather plaie with shadowes, than bestow their wits about profitable and serious matters .... If any be desirous to heare of their charming qualities, wherewith the Circes and magicians haue vsed to bring to passe their diabolicall incantations, let them read *Homer* touching that matter in the twentie chapter of his *Odysses*, and there shall they finde matter scarce worth the reading. (145)

Homer's poetic account of the plant may be "scarce worth the reading," but Gerard nevertheless cannot resist directing his reader to it with a remarkably precise reference. Gerard shows a similar sort of reluctant skepticism about what he considers to be a fictional account, this time offered not by poets but by several modern herbalists: "*Ruellius* and such others haue reported many strange tales hereof, scarse worth the noting, saying that the seede of flaxe put into a radish roote or sea Onion, and so set doth bring foorth that herbe Tarragon" (193). As in his account of the poetical fiction about Moly, here too Gerard is unable to resist including, paraliptically, the fiction he denounces. Likewise, in his discussion "Of the Sorrowfull tree, or Indian Mourner," he writes that its flowers, which bloom only at night,

> in the day time looke withered and with a mourning cheere: the leaues also at that time shrinke in themselues togither ... very sadly lumping, lowring, and hanging downe the head, as though it lothed the light, and could not abide the heate of the sunne. I should but in vaine lose labour in repeating a foolish fansie of the Poeticall Indians, who woulde make fooles beleeue, that this tree was once a faire daughter of a great Lord or King, and that the sunne was in loue with hir; with other toies which I omit. (1343)

Of course, he has done anything but omit the "foolish fansie."[12] Moreover, in his description of the plant, which interprets physiology as disposition, his prose participates eagerly in the personification which appears to derive, like the plant's name, from this very tale.

These examples show that, despite the suggestion that "Gerard showed greater fondness for lore and fanciful stories that [*sic*] the plain truth" (Hoeniger, *Medicine* 47), his attitude was consistently less consistent than that. Instead, he is conflicted about the place of fiction in his herbal, while at other times he shows little awareness of switching from a factual to a fictional register and sometimes back again. For instance, his explanation of the etymology of a plant called Amaracus sways perceptibly back and forth between being presented as an account Gerard believes and a tale he recognizes to be fiction:

> The cause of the name of this most sweete and pleasant herbe is not determined except it came through the faining of the Poets, who report that those of Cypres fawning vpon their king *Cinara*, imagining to please his humor, said that his sonne in time of his youth, carried a boxe full of fragrant ointment, ouer the

[12] In describing Mugwoort he makes a similar rhetorical move, allowing him to have his "fansie" and mock it too: "*Pliny* saith, that the trauailer or wayfaring man, that hath the herbe tied about him, feeleth no wearisomnes at all, and that he who hath it about him can be hurt by no poisonsome medicines, or by any wilde beast neither yet by the sunne it selfe; and also that it is drunke against *Opium*, or the iuice of blacke Poppie. Many other fantasticall deuises inuented by Poets are to be seene in the works of the auncient writers, tending to witchcraft and sorcerie, and the great dishonor of God: wherefore I do of purpose omit them as things vnwoorthie of my recording or your reuiewing" (946).

> fieldes of most pleasant herbes, which by mischance he spilt vpon the said herbes, which being moistened therewith, yeelde vnto this day that excellent sauour, wherein we do so much delight. This boy mourning for the losse of his ointment, the gods (as the poets faine) in consideration of his parentage and excellent perfection, did change and transforme the boy into that herbe, which is called *Amaracus*, after his owne name. (539)

His twice-repeated reference to the "faining of the Poets" seems evidence of his firmly skeptical attitude to the verity of the tale. But other details belie this awareness, such as his assertion that the plants "yeelde vnto this day that excellent sauour, wherein we do so much delight"—apparently, a very real "sauour," one witnessed and enjoyed by his own contemporaries, that was somehow produced by the spilling of an imaginary ointment. Elsewhere Gerard shows what seems a genuine indifference to the mingling of fiction and fact: "The fragrant smell that this kinde of *Ambrosia*, or Oke of Cappadocia yeeldeth, hath mooued the Poets to suppose that this herbe was meate and foode for the gods: *Dioscorides* saith it groweth three handfuls high; in my garden it groweth to the heigh of two cubits ..." (950). This appears to be the most compact example of Gerard's catholicity when it comes to standards of evidence: he refers first to poets, then to the chief of the ancient herbalists, and finally to measurable and empirical eye-witnessed evidence of his own experience—yet he does not distinguish among these authorities in terms of their reliability. Indeed, the sheer adjacency of the claims, without even a qualifying conjunction between them, suggests that all such evidence is to be weighed equally. This implies not so much that Gerard had a greater fondness for lore than for plain truth, but that the standards for distinguishing between those categories and among types of evidence were not especially apparent or important to him.

At other times, however, Gerard voices nothing but disgust for those who promulgate botanical fictions.

> The bright shining red Beares eare of *Mathiolus* description, seemeth to late Herbarists to be rather a figure made by conceit or imagination, then by the sight of the plant it selfe, for doubtlesse we are perswaded that there is no such plant, but only a figure foisted for ostentations sake, the description whereof we leaue to a further consideration, because we haue not seene any such plant, neither do we beleeue there is any such. (641)

But Gerard then blithely includes a figure of the figment on the next page. Nor is this the only time that he includes images of fictional or otherwise dubious plants in the herbal, only to dispute their existence in his text. Chapter 38 of Book One, for instance, opens with two illustrations set side by side: the first is labelled "*Zinziberis fictum.* The fained figure of Ginger" and the second "*Zinziberis verior Icon.* The true figure of Ginger" (54). Gerard has, in other words, purposely included a fictitious species of plant in his herbal, laying out its illustration in exactly the same format as all the non-imaginary species that he portrays. It is true that the caption reveals it to be fictional; but since the caption appears both in Latin and in English, and in the same format as all the other names, it still contributes to the figure's camouflage as a genuine plant.[13]

In a chapter entitled "Of two fained pictures," Gerard describes his motivations for dedicating an entire chapter, including illustrations and descriptions, to two presumably fictional plants:

> I haue thought it conuenient to conclude this historie of the Hyacinthes with these two Bulbus plants, receiued by tradition from others, though generally holden for fained and adulterine. These pictures I could willingly haue omitted in this historie, if the curious eie could else where haue found them drawn & described in our English toong: but because I finde them in none, I wil lay them down here to the end that it may serue for excuse to others who shal com after, which list not to describe them, being as I said, condemned for fained and adulterine ... (107)

Gerard includes the fictional plants, all but paradoxically, so that no one else need do so again. He goes on to discuss their names and give a detailed description of the physical features of both, building up vivid images that can only add to an appearance of thoroughgoing verity in the reader's mind; but he ends with a conclusive re-assertion of their fictitious natures: "The virtues and temperature are not to be spoken of, considering that we assuredly perswade our selues that ther are no such plants, but mere fictions and devises" (107–8). Why it is more possible or appropriate to discuss the physical features of a fictional plant than its "vertues" is not explained. Gerard's complex motivations for including non-existent plants in his herbal are further elaborated in the chapter on various kinds of Mountaine Sanicle; when he reaches the fourth kind, he writes,

---

13 These are not the only examples of Gerard's willingness to include a plant whose existence he denies, and thus to introduce a kind of fiction into an account he otherwise often strives to portray as factual. Elsewhere two figures are labeled, in English, "Beares eare Sanicle" and "Beares eare Sanicle the true figure" (645). A reader who only glanced at the first would be forgiven for thinking that it is equally "true"—since it is not labeled "false" or "fained" to distinguish it. Another instance also involves two images set side by side, one labelled Yellow bulbed Flower de-luce, the other Ash Colourde Flower de-luce, and neither identified as true or false; yet of the latter, he writes, "This pale yellow ash coloured bulbus Flower de-Luce (if there be any such) agreeth with the former in description. I say if there be any such; for in mine opinion there is none such differing from the other" (93). His opinion is clearly that the Ash Colourde Flower de-luce is a fiction, yet he includes not only the picture but space for a separate description of it. Even these two aspects of page layout seem to imply the factuality of Ash Colourde Flower de-luce; but Gerard then uses the space set aside for its description only to assimilate its appearance to that offered in the previous description and to deny the plant's existence. While Gerard comes down conclusively against the existence of the Ash Colourde Flower de-luce, his dedication of both pictorial and textual space to its description suggests he has, at the very least, an odd relationship with counter-factual botanical matters.

> There is another figure set foorth which may seeme vnto the ignorant to be one of the same kinde, but differing in some points, whereas in truth they are but one and the selfe same plant; one of the which I would willingly haue left foorth of our volume, but fearing that some not knowing that they be one, might aske the cause why we left it out; but hencefoorth I doubt not but those that shall succeede will leaue it out as needlesse, considering they know that one picture was made by a drie and withered plant, and the other by the liuely herbe it selfe. (645)

As before, Gerard takes up the burden of including a fictional figure so that others need not bother do so in future, thus confirming his vision of his herbal as a provisional draft in a larger and collaboratively authored Book of Nature.

As with such images, Gerard is also not averse to the inclusion of stories he does not believe; he might even present them without any introductory signal of their fictive nature. Here is some of what he writes, and even cites, with a straight face:

> This *Aglaophotis* of the earth, or *Cynosplastus,* is called of *Iosephus* the writer of the Iewes war, in his seuenth booke 25. chapter[,] *Baaras*, of the place wherein it is found; which thing is plaine to him that co[n]ferreth these things which *Aelianus* hath written of *Aglaophotis* of the earth, or *Cynospastus*, with those which Iosephus hath set down of *Baaras:* for Aelianus saith, that *Cynospastus* is not plucked vp without danger; and that it is reported how he that first touched it, not knowing the nature thereof, perished. Therfore a string must be fastned to it in the night, and a hungrie dog tied thereto, who being allured by the smell of rosted flesh set towards him, may plucke it vp by the rootes. *Iosephus* also writeth, that *Baara* doth shine in the euening like the day star, and that they who come neerre, and would plucke it vp, can hardly do it, except that either a womans vrine, or hir menses be powred vpon it, and that so it may be pluckt vp at the length.
>
> Moreouer, it is set down by the said author, as also by *Plinie* and *Theophrastus*, that of necessitie it must be gathered in the night; for if any man shall plucke of the fruit in the day time, being seene of the Woodpecker, he is in danger to lose his eies; and if he cut the roote, it is a chaunce if his fundament fall not out.

Only at this point does Gerard write off all he has written thus far as mere fiction:

> The like fabulous tale hath beene set foorth of Mandrake, the which I haue partly touched in the same Chapter. But all these things be most vaine and friuolous: for the roote of Peionie, as also the Mandrake, may be remooued at any time of the yeere, day or hower whatsouer.

The moral of the story is then leaned on rather heavily, as if to justify the lengthy telling of the tale:

> But it is no maruell, that such kindes of trifles, and most superstitious and wicked ceremonies are found in the bookes of the most ancient authors; for there were many things in their time very vainly fained and cogged in for ostentation sake,

> as by the Aegyptians and other counterfeit mates, as *Plinie* doth truly testifie, an imitator of whom in times past, was one *Andreas* a Phisition, who as *Galen* saith, conueied into the art of physicke lies, and subtile delusions. For which cause *Galen* commanded his scholars to refraine from the reading of him, and of all such like lying and deceitfull sycophants. (832).

Gerard's ambivalence toward the place of fiction in herbalism is perhaps at its height in this anecdote. Of particular interest is the order in which he presents his information. He recounts in full detail bizarre stories about the difficulty of picking peonies, citing multiple sources—including those generally respectable ancients, Pliny and Theophrastus. Not until he has finished the story does he make known his judgment: that it is but a "fabulous tale" and that "all these things be most vaine and friuolous." He follows up with a paragraph justifying the inclusion of "such kindes of trifles" in "the bookes of the most ancient authors" on the grounds that his classical predecessors had a kind of cultural blind spot about such things "for ostentation sake." He notes that one of these ancients finally rose above such tales and "commanded his scholars to refraine from the reading" of them—yet Gerard himself happily includes them for his own readers to peruse. He exhibits a similarly paradoxical attitude in relation to another plant of which fabulous tales have been told:

> But this is to be reckoned among the old wiues fables, and that also which *Dioscorides* telleth of, touching the gathering of Spleenewoort in the night, and other most vaine things, which are founde here and there scattered in the old writers books: from which most of the later writers do not abstaine, who many times fill vp their pages with lies and friuolous toies, and by so doing do not a little deceiue yoong students. (979)

He condemns other "later writers" for including in their books the very "lies and friuolous toies" that he has just finished including himself.

Gerard's botanical work has been embroiled in bibliographic controversy from the beginning. Whether his authorship was a legitimate form of anthologizing, borrowed from the larger literary culture of commonplacing, or whether he was simply plagiarizing will never be resolved. Moreover, the argument against his sole and original authorship starts to seem moot when the way in which botanical authorship functioned in the period is examined more closely. Like poems in a garden of verse, a period understanding of plants was always gathered from many sources: from anonymous and named poets, ancient and modern, as well as from both ancient and modern herbalists; from sometimes acknowledged but often unnamed women and husbandmen; from servants sent to collect plants from abroad, and from gardeners who sent plants and information by correspondence. Gerard's own contributions to his herbal were also more diverse than the term "author" seems to imply: he was an author, but also a collector and harvester of other gardens, texts, and the common stock of knowledge harboured among husbandmen and herbwives. He was both a fastidious ascertainer of facts and a rampant promulgator of fictions, wittingly and otherwise. His approach to naming plants demonstrates the interpenetration of an authorial and an anthological mindset: even in the case of a name he himself made up (like "*Robinus* gum Succorie" (224)), he cites within the fabric of the name the authority of another herbalist. Elsewhere, the infolding of citations among poets and herbalists becomes so complex as to blur the line between the two kinds of writing; so does Gerard's aesthetic evaluation of his subject matter, and his development of a kind of an artful botanical poetics as part of his rhetoric of description. Gerard's herbal is thus more proximate to the poetry of the period than one might expect: in terms of his expressed aesthetic values, his rhetorical strategies, his citation of poetic texts, and perhaps most importantly, his anthological approach to authorship.

Chapter 5
Domesticated Plants and Domesticating Books: Cultivating Household Textual Collection

Well before the sixteenth century, plants in England had long held a diverse array of functions in households crossing the bounds of social rank; books, on the other hand, had not. In her study of the economic history of the English book trade, Marjorie Plant notes that "in the sixteenth century the buying of a book was something of an event in a man's life, calling for as much forethought as the buying of a piece of furniture for his house" (206), and Sears Jayne has argued that "outside the Universities very few people in England before 1590 owned as many as fifteen different books" (14). Moreover, and especially in the early part of the century, participation in book culture could be attended by very real personal risks.[1] As the century progressed, however, and print became more prevalent, books increasingly did find their way into homes as part of the process of what Roger Chartier has characterized as "typographical acculturation" (159). Peter Clark's study of almost 3,000 household inventories shows that book-owning rose fourfold in three Kentish towns between 1560 and 1640. Books clearly and increasingly crossed the domestic threshold—but then where were they put? Where were they used? Although books may have been more and more present in the homes of individual owners, books lacked an entrenched and meaningful place there: that is, they lacked both a physical site devoted to their storage, display, and use and a social status that underwrote their occupation of that site. This was especially true of printed, vernacular, secular works, and more especially of the miscellaneous kind so often characterized as gardens of verse. Of course, their makers (both authors and printers) could only hope they would gain such a place, so that buyers would not hesitate at the bookstall to buy them and to bring them home.

One reason potential book-buyers might hesitate is that they would not accrue personal status through the purchase of most printed vernacular books, as they might

[1] A.E.B. Coldiron has recently instantiated some of these risks in the first part of the sixteenth century: "In 1524 the Bishop of London warned booksellers not to import heretical books; Wolsey's book burning in 1525 also meant that no new book was to be printed without his specific authorization. In that same year the translation of the *Ymage of Loue* (trans. J. Gough and printed de Worde) was investigated. In 1526, T. Berthelet's translations of Erasmus were investigated. In 1531, Richard Brayfield was arrested and later burned for importing foreign prohibited books" (222).

with more learned volumes, with expensive manuscripts, or with other commodities altogether. When inventories of books were made, appraisers tended to glance past the less valuable English books in collections (Peter Murray Jones 56). Often vernacular works, which tended to encompass the more ephemeral genres, are grouped without their titles at the end of such lists, if they appear at all (Plant 263). One scholar of early modern libraries notes that printed work, in particular, "did not at first find a ready acceptance among the wealthy bibliophiles" (Kibre 258). And when such books were collected, they were not necessarily respected; William Webbe, in his *Discourse of English Poetrie* of 1586, condemns the collection of "innumerable sortes of Englyshe Bookes, and infinite fardles of printed pamphlets" with which, he claims, "every study [is] stuffed" (Aiiii). In this instance, vernacular printed books are at least imagined to be present in the home, and specifically in the study—but their presence is only considered grounds for contempt. The studies he mentions were, it seems rare enough, since even the educated Elizabethan tended not to have a separate room set aside as a library or even a private study or closet, the latter relatively rare if rapidly emerging household spaces. Where such a room did exist, it was very unlikely to contain an entire wall lined with bookshelves, let alone bookcases. This is a far cry from the respectable role assumed by the domestic library in early eighteenth-century England, as is evidenced by Joseph Addison's description of his encounter in 1711 with a private library filled out in an appearance of plenitude with "Counterfeit Books upon the upper Shelves, which were carved in Wood, and served only to fill up the Number" (Addison and Steele 146; see also Birrell 128–9). Such libraries did not exist a century before; Schofield notes that "books did not merit a special room in medieval or Tudor London houses, though they did by the end of the seventeenth century" (309–10), by which time Samuel Pepys had famously fitted up a room full of custom-made cases to contain exactly 3,000 finely-bound books, numbered and arranged on the shelves from smallest to largest, with carefully measured wedges placed beneath individual books to produce a uniform height in each row (Petroski 126–7). His library was not just a collection of favourite tomes; it was, like so many early modern collections of *naturalia* and *artificialia*, a visible order that represented "an attempt to create a universe of knowledge, a replica of the Divine Book of Nature in which every gap might be filled, every subject accounted for" (Boesky 312). No Elizabethan of similar status is known to have had anything remotely comparable. Between these two periods, what tactics were undertaken—by writers, printers, and readers—to allow collections of books to become so respectably domesticated? One such tactic may be found in the pervasive botanical metaphors identifying early modern books. When books were likened by their makers to collections of plants, the allusion could be read as staking a claim with regard to the meaning, value, and desirable place of books in the home.

The later-sixteenth century in England was a time when not just title pages but the very titles of books "became not only a way of identifying and describing texts but also a procedure for advertising and disseminating works in the public arena" (Shevlin 47; see also Saenger). Such marketing was a task peculiar to printers, whose profit depended on selling a significant number of a single title:

"With no precedent for how to market and publicize their product, publishers were faced with inventing ways to advertise their wares. Having long acted as the tool that identified texts, the title was ideally situated to induce demand by assuming the additional functions of announcing, describing, and even hawking a work" (Shevlin 48). Printers could increase profits by finding ways to let books gain a greater foothold in domestic space. Tessa Watt notes that "[t]he idea of making a paper imitation of a physical object ... seems to be symptomatic of the early phases of print, when books were never just books, but 'looking glasses', 'maps', 'posies' and so on" (248); she speculates that "in a society adjusting to literacy, print was not just print, but was incorporated into daily life by allying itself with other methods of communication" (251). Texts troped as gardens, garlands, or groves may have improved their chances of being brought into sixteenth-century English houses by ascribing to themselves the established values of plants that were already present in the same spaces. Moreover, some conventional domestic places for plants were at least metaphorically and sometimes materially appropriated by books in the later sixteenth century. Such premises as these shall allow a survey in this chapter of the connections between the relatively novel practice of private book-collecting and the historical conventions of collecting, storing, and displaying botanical materials in the home.

The Place of Books

With the rise of print and more widespread consumer culture, books increasingly found themselves in the hands of private owners in late sixteenth-century England, a situation that made new demands on the organization of household space at a time when a considerable cross-section of residential space was being reorganized for other reasons in the course of what has been called and become known as the Great Rebuilding (Hoskins; Beard 26). Such reorganization opened up new ways of using both books and the spaces in which they came to be kept. A new or prospective owner of books in this period would need to face questions to do with the place of books in the domestic economy: where was one to keep them—on display or in storage? Who should see them and use them? Where, when, and how was it appropriate to read them? Such questions confronted the potential buyer of books both at and after the point of sale, since "Elizabethans did not buy their books ready bound, and hard covers compete[d] with walls as the proper place to store or display at least some kinds of written text" (Fleming, "Wounded Walls" 8). Decisions had to be made.

These were new decisions for many since, until this period, books had been almost exclusively luxury items for the very rich. Before moveable type, even the wealthy or scholarly might possess only a handful of elaborately-bound works; and even "[l]ong after the invention of printing, books continued to be regarded as so great a luxury that in the majority of households a chest, hutch, or shelf beneath a desk sufficed to contain them" (MacQuoid 78). Larger collections were mainly housed in ecclesiastical and educational institutions, where their value was such that they were often secured by chains to shelves and lecterns (Streeter). Bookcases, even in the sixteenth century, were found only in institutional collections; they were simply not an aspect of domestic furnishing (Edwards 51). Until the later sixteenth century, then, the average, educated English homeowner was not likely to own many books, while any he did have would generally be locked in a chest or stored beneath the sloping lid of a desk, out of sight and often out of mind (Clark, *Libraries* 48–9).

Cheap print began to alter this state of affairs; moreover, the dissolution of ecclesiastical libraries starting in the 1530s transferred thousands of valuable books to private hands (Jayne 39–43). In response to this unprecedented abundance, people came up with what a historian of the bookshelf calls "a staggering number of ... increasingly ingenious devices for holding and storing books"—so many and so varied in invention that the same writer concludes that "questions of where and how to work with books in carrels, private studies, and libraries had come to be a serious topic of thought, conversation, and construction" (Petroski 103). Evidence of such concern in later sixteenth-century England about the keeping of books may be found in the will of Thomas Lorkyn, Regius Professor of Physic at Cambridge—a man who, as a result of his occupation, had a much larger collection of books than his average contemporary. Lorkyn left his library of approximately 400 medical books to the University of Cambridge, but made his bequest contingent on the university coming up with a suitable storage system; he left "all my phisicke bookes with thos that appertaine allsoe to phisicke to be kept in the inward Librarie in a gret Cubbord locked with two locks the librarie keeper to have on key and the phisicke Reader an other for him and his auditors. Students of phisicke to looke one soe that the said university be at chardge to make convenient cubbords or presses or Cubbord or presse to keepe them in Locked within on yeare after my death" (qtd. in Peter Murray Jones 167). If the university did not come up with such a system in that time, his books were destined for Peterhouse instead.

Yet few similar conversations and fewer such constructions have descended intact to the present day. Peter Clark has noted that there is very little evidence about "where books were usually kept" (103) in this period. This is true even of the greatest private libraries: Lords Lumley and Burghley kept collections of more than one thousand books, yet no one can determine "just how or where" they kept them (Girouard 164–5). West notes that after the monastic dispersals, only "the secular, private libraries ... must have supported the English Renaissance," yet such libraries "are barely present in the research literature" and "the rooms themselves seem lost" (71). It was only as late as the second half of the seventeenth century that cases of books start to form a "prominent and fixed part of a unified scheme of decoration" in English homes (Peter Thornton 314); since so many analyses of the subject begin in the later seventeenth century, the private English library sometimes appears to spring to life fully formed and furnished at this time and thus "appear[s] to have no history" (West 71). In what follows I would like to try to alter that appearance, despite the difficulties presented by a dearth of direct evidence, by examining how the novel activity of domestic book collection appeared to pattern itself on the more established botanical collections that occupied various household spaces, from gardens to galleries and stillrooms in private closets. I suggest that one of the ways in which producers expressed this patterning was by reimagining conventional metaphors of books as plants in light of a contemporary culture of domestic plant use.

Literary historians have only recently begun to address books as socialized material objects, a task that necessitates remembering the places in which they were kept and used. Library scholars, on the other hand, have tended to concentrate on reconstructing institutional examples of English Renaissance libraries and their furnishings, perhaps because it is in cathedral and college libraries that the best evidence remains. As West notes, the domestic reading room "appears to have fallen through the gaps between research communities, and is a missing element in the general move towards a social history of books" and the history of reading (68). Yet this kind of reading space is especially integral to literary history, if only because so many of the early modern books characterized by botanical metaphors contain the type of text that literary history has traditionally favoured: poetry and fictional prose. These texts, however, equally tend not to be canonical epics and odes so much as what literature is when it's at home: the stuff of bedside tables, parlours, nurseries, and private studies. If there is a genus of text whose natural habitat is the domestic interior, its species may have been studied in considerable detail, but its early modern habitat is less well-known. While much empirical evidence may be beyond recovery, other approaches to the topic are at hand. West has begun to forge new paths by suggesting "wider research questions about how books and people interact, through the medium of libraries," including questions about "the material context" of books that can contribute to an understanding of "the social life of the built space" (West 69–70). This chapter is meant to contribute to this aspect of the material and social history of books and reading by expanding the material context to include a consideration of its metaphorical extensions, as Renaissance readers, writers, and printers did; as Watt put it, early modern books were never just books (248).

The Place of Plants

In the sixteenth century, the relatively new but increasingly widespread recreational or pleasure garden overlapped conceptually and materially with more customary medicinal and culinary gardens; in each case, plants were grown expressly to be displayed, stored, and consumed inside the house, supplying much of its food, drink, medicine, and ornament. Historians posit a plethora of cut and growing flowers, fruit, branches, and herbs in vases, pots, hearths, and window boxes of private homes, as well as on floors, in matted or strewn form (Amherst 162–3; Thomas 224; Blacker 32). The Dutch physician Levinus Lemnius wrote after visiting England in 1560 that the "Chambers and Parlours strawed ouer with sweete herbes, refreshed mee, their Nosegayes finely entermingled with sundry sortes of fragraunt floures in their bed cha[m]bers and priuy roomes, with comfortable smell cheered mee up and entirelye delyghted all my Senses" (Thomas Newton 48). Lemnius also wrote that "no Nation more dece[n]tly, more trimmely, nor more sightly" strewed its floors with sedge or decorated its parlours with "green boughes, fresh herbes or vine leaues ... then they doe in Englande" (48). The prevalence of plants in English houses seems to have been particularly remarkable from a comparative cultural perspective.

Those writing prescriptively also made recommendations in this direction: Thomas Tusser's *Fiue hundreth points of good husbandry vnited to as many of good huswiferie* (1573) includes one list of "Strowing herbes of al sorts" (39[r]) and another list of "Herbes, branchis & flowers for windowes and potts," including red and white carnations set in spring and fall "in potts, payles or tubs" (40[v]). The extent to which flowers grown out of doors were brought inside houses for decorative purposes may be evidenced by the frequency with which Gerard states in his herbal that the only reason a particular plant is grown in gardens is "for the flowers sake, to the decking vp of houses, for which purpose it chiefly serueth" (*Herball* 360). He also speaks of houseplants like "Herbe Aloë" being "hanged on the seelings and vpper posts of dining roomes" (409). Similarly, in a section of his *Floraes Paradise* entitled "A Garden within doores," Hugh Plat suggests the creation of a "faire gallery, great chamber or other lodging ... to be inwardly garnished with sweet hearbs and flowers, yea & fruit if it were possible" by means of pots arranged "to stand loosely vppon faire shelues" that may be transported in and out of the house "with a pulley from your Chamber windowe into your Garden" (31–2). Plat also suggests that "[i]n euery windowe you may make square frames" and "fill them with some rich earth, & plant such flowers or hearbes therein, as you like best" (32); he even goes so far as proposing the growing of barley and other plants within cucumber plants that are in turn to be suspended from the ceiling (33–4), and vines that travel from the outer walls into the domestic interior to "run about the sides of your windowes, and all ouer the seeling of your roomes" (36). Moreover, as a few excerpts from *The Grete Herbal* (1529) make clear, plants were thought to contribute to the psychological or social atmosphere of a house and its inhabitants:

> 'To make folke mery at ye table', we are told to 'take foure leves and foure rotes of vervayne in wyne, than spryncle the wyne all about ye hous where the eatynge is and they shall be all mery.' The smoke of 'Aristologia ... dryveth all devyllsshnesse and all trouble out of the house' .... [T]he lesser-mugwort should be laid under the door of the house, for, if this is done, 'man nor woman can not anoy in that house' (Arber, *Herbals* 47).

Those in the plant business thus recommended that plants be brought inside the house for both aesthetic and utilitarian purposes, and ideally for a combination of the two. In this way, plants made the entire interior of the house their naturalized habitat in the sixteenth century.

One aspect of the botanical tropes applied to books that is not often noticed is that books characterized as gatherings of plants—as gardens or groves, arbours or bowers—are often also characterized as interiors that bore close and working relations with the domestic interior. In a study of early modern research libraries, William Clark has noted the conceptual overlap of different interiors for institutional and individual collections in the period:

> Into the Baroque era, colleges and universities, like private collectors with their *Wunderkammern*, did not consistently distinguish between libraries, museums, cabinets, laboratories, and often not between those and archives and treasuries. For a good part of the early-modern era, spaces we tend to separate were commonly not. (190)

One similar space that Clark's list overlooks is the garden. The oversight may arise from the fact that the garden is less obviously than the other spaces that he names a space that is bounded by walls, a basic criterion for the site of most collections. Yet the early modern garden, and especially the Tudor garden, was a definitively enclosed space. In the sixteenth century, "the idea of such a thing as a practically unenclosed garden had not, as yet, entered men's minds" (Amherst 112–13), while as late as 1629, John Parkinson advises that a garden should have "helpes of bricke or stone walls to defend it" (*Paradisi* 1). Such walls enclosed collections of often highly valuable botanical artifacts, ranging from medicinal herbs in traditional physic gardens to expensive new bulbs in the increasingly prevalent elite recreational gardens. The identification of books with gardens can be related to the way actual domestic gardens functioned as, and in relation to, the larger domestic interiors to which they belonged.

John Gerard writes in the dedication to his *Herball* of "the great charge and care of almost all men in planting and maintaining of gardens, not as ornaments onely, but as a necessarie prouision also to their houses" (A2ᵛ). His characterization of the domestic garden as both ornament and necessity epitomizes the way such gardens embodied the fashionable Horatian code by being both delightful and instructive, containing both profit and pleasure, things otherwise so often at odds (Beretta 15). Yet the Horatian formula of *dulce et utile* was not intended for gardens, but for poems—and was ubiquitously observed in the textual collections troped as plants. See, for instance, *The garden of pleasure: containing most pleasant tales, worthy deedes, and wittie sayings of noble princes [et] learned philosophers, with their morals*, which bills itself on its title page as "No lesse delectable, than profitable" (Guicciardini), or *The floures of philosophie with the pleasures of poetrie annexed vnto them*, which similarly asserts up front that it is "as wel pleasant to be read, as profytable to be followed of all men" (Plat). The subtitle of *The saints nosegay* paratactically links the ability to combine pleasure and profit to salient botanical qualities when it identifies its contents as *A posie of 741 spirituall flowers both fragrant and fruitfull, pleasant and profitable* (Clarke). Books identified with gardens, as these examples demonstrate, explicitly identified with the garden's ability to combine pleasure and profit; the figure of the book as garden may thus be read as staking a claim for its combined usefulness and fitness for display in the household spaces where the garden and its produce were already found.

To claim a place in the garden, moreover, was to stop just shy of claiming a place in the house itself. The Tudor garden was closely integrated with the domestic interior, and even resembled an interior itself. The continental theories of such writers as Leon Battista Alberti—who instructed that the recreational garden was to be treated as structurally, even axially, integral to the design of the house (Hobhouse 140)—appeared in English thought and practice in the second half of the sixteenth century and especially in its final quarter (Beretta 66, 78). Moreover, native tradition required contiguity between house and garden for purely practical reasons: adjacency facilitated the security, cultivation, and collection, inside the house of valuable horticultural contents (Bushnell, *Culture* 141). Consequently, gardens "came to be regarded as attached to a house" (Quest-Ritson 22). The garden was a kind of threshold between public outdoor space and more private indoor space; but its placement, usually at the back of a building or in a courtyard, meant that it was more emphatically private than the front gardens of later periods.

Within a Tudor garden's walls were most likely to be found separate, boxy beds which were often further subdivided in linear designs. One historian of Renaissance gardens in England describes them as "a series of enclosed squares and rectangles, separate visual experiences bounded by walls and hedges" (Strong, *Renaissance* 60). The idea of the garden as a subdivided architectural space was further developed through the arbours, summerhouses, and banqueting houses of the more elaborate sixteenth-century pleasure gardens.[2] Because of the tendency to parse a garden into separate architectural units, "the garden visitor's eye was meant to be drawn to the details of the parts, to the individual plants and the elaborate knotted beds, rather than to the design of the whole perceived in a vista or view"; walking among such beds and buildings may thus have felt much like walking within the halls of its attached house, since in the garden, too, "[t]he visitor passed from one garden 'room' to another" (Bushnell, *Culture* 136, 138). In these rooms or beds, plantings tended toward the sparse and widely spaced arrangements that we might be inclined to associate today with a botanical garden (Strong, *Artist* 35). The sum effect of the garden of the period was of an arrangement of separate spaces featuring distinctly separate items, with the whole array of smaller enclosures itself enclosed within the property lines of the house. The experience of such a garden would also, however, have been analogous to that of a garden of verse. Such books often contain short or fragmentary texts with stanzas of verse that appear visually on the page as distinct rectangular figures on a larger rectangular ground. Such texts also provide "separate visual experiences"—as opposed, for example, to the uninterrupted visual experience

[2] In 1598, John Stow wrote in his *Suruay of London* of the proliferation of city gardens "wherein are builded many fayre Sommer houses ... with Towers, Turrets, and Chimney tops, not so much for use, or profites, as for shew and pleasure" (353). Summerhouses or banqueting houses were often erected temporarily in the gardens of the wealthy to mark a single special occasion (Amherst 116). These would frequently be built of or decorated with botanical materials (see Parkinson, *Paradisi* 610 and Gerard, *Herball* 746).

enabled by reading continuous prose or even longer narrative verse. The visual experiences of the Tudor garden are separate because they are "bounded by walls and hedges," just as the visual experience of the garden of verse is bounded by the edges of the pages and sometimes further by the printer's decorative rectangular fleurons bordering those pages, as in the continuous border surrounding each page of text in Anne Wheathill's *A handfull of holesome (though homelie) hearbs,* the design of which closely resembles the borders of contemporaneous knot gardens. Just as a visitor passes "from one garden 'room' to another," a reader turns from one text to the next in a miscellany, a genre and a form in which skimming and skipping are encouraged since the linear arrangement of the texts is often only an accident of the medium.[3] The Tudor garden was designed to be experienced, then, as textual compilations were designed to be read: for the pieces and parts that stood out and signified to the reader on any particular occasion of reading.

That early modern reading was distinctly more discontinuous and disjunctive than later and more linear approaches has been argued variously by Roger Chartier, Lisa Jardine and Anthony Grafton; Mary Fissell summarizes their general hypothesis that such "reading depended upon short items which were not read as part of a narrative sequence, but were made sense of individually" and possibly "combined in various satisfying ways" (78).[4] One notably disjunctive genre, intended for consultation and comparison rather than through-reading, was the herbal, with its short, self-contained entries, headings and sub-headings, and overall absence of narrative; the medicinal simples described were, likewise, often intended to be "combined in various satisfying ways" in pharmaceutical recipes. The discontinuous experience of the Tudor garden—with its separate beds and sparse plantings—correlated remarkably well to the discontinuous reading experience of texts explicitly related to the garden's contents; the form and experience of the garden were reflected in the form and experience of the herbal. Non-herbal textual collections that offered similarly discontinuous experiences could thus allude to valuable cultural precedents, both in other kinds of books about plants and in actual botanical collections.

Gallery and Paradise

The contents of sixteenth-century gardens were frequently collected and kept, for display or use, inside the house; but another and less obvious way in which

---

[3] With respect to Tottel's miscellany, Elizabeth Pomeroy writes that "[h]e evidently gave little thought to progression or shape in the volume as a whole, an indifference shared by the compilers of most later miscellanies" (35); she later refers, in this monograph on Elizabethan miscellanies, to their "random structure" (118).

[4] A recent analysis of the balance of evidence for discontinuous versus continuous reading and writing practices in Tudor England cautiously concludes that "the balance ... inclined to the side of fragmentary reading, with through-reading reserved for special needs" (Lethbridge 73).

the garden was present in the domestic interior involved the visual and often olfactory relations of certain rooms to the garden, relations especially enabled by the presence of windows. This is an aspect of domestic history that is often overlooked by architectural historians, who "rarely say much about the effect of windows on the interior" (Peter Thornton 80). Yet such spectacles were not incidental but deliberate, with the largest windows often placed in upper-storey rooms specifically to permit the best view of the garden, which was in turn often designed to be seen from a bird's-eye view (De Bray 86). Gervase Markham, writing in *The English Husbandman* in 1613 about the ideal design of "a plain countryman's house," situates the garden with at least as much attention to this visual relation as to more horticultural concerns: "on the South side of your house, you shall plant your garden and orchard ... for the prospect thereof to all your best rooms" (214). Parkinson echoes this sentiment, making it the first order of business in his gardening book of 1629 when he writes that the ideal situation involves "the fairest buildings of the house facing the garden" because "the buildings and roomes abutting thereon, shall haue ... the beautifull prospect into it, and haue both sight and sent of whatsoeuer is excellent, and worthy to giue content out from it, which is one of the greatest pleasures a garden can yeeld his Master" (1). The pleasurable visual experience from on high of intricately patterned gardens may be seen as analogous to the reading of a book—particularly of an anthology, the kind of book that lets a reader survey a field (to revive a deadened metaphor). One image in Thomas Hill's *The Gardener's Labyrinth*, a practical handbook, shows a garden that bears a notable resemblance to an open book (Figure 5.1). According to Hugh Plat in his *Floraes Paradise*, topiary forms in the shape of "pyramids, losinges, circles, [and] pentagons" could be "planted with herbes" that would "grow greene, according to the forme they are planted in" (68–9); the very same geometric shapes are recommended as forms for poems in George Puttenham's *The Arte of Poesie*. Puttenham discusses forms that "yeld ... an ocular representation, your meeters being by good symmetrie reduced into certaine Geometricall figures" such as "The Spire or taper, called piramis," "The Lozange," "The Rondel or Sphere," and "The Lozange rabbated," a pentagonal form (104–6). A visual perception of the garden was thus susceptible to comparison with the visual perception of a poetic text in its material form, as lines and letters shaped on a larger rectilinear ground.

The upper-storey long gallery, a novel and increasingly chief room of great houses in the sixteenth century, especially enabled such a prospect on the garden below (Cooper 303; Burton and Kelly, *Elizabethans* 57–8). In London, as likely elsewhere, "the gallery always bordered, or ended with a view of, the private garden" (Schofield 310); such a relation between gallery and garden is implied in the title of *A Goodly Gallerye with a most Pleasaunt Prospect, into the garden of naturall contemplation* (Fulke). The large glass windows of the gallery allowed the entry of brightness and the circulation of fresh air and floral fragrances, some of the most valued qualities of the outdoors.[5] As William Lawson put it in his

[5] See, for example, the discussion of the medicinal value of florally-scented air by the physician Levinus Lemnius (Thomas Newton 47–8).

Fig. 5.1 An illustration from Thomas Hill's *The Gardener's Labyrinth*

New Orchard and Garden, a horticultural manual published in 1618 that reflected his experience of the previous half-century: "it is (no doubt) a comfort to them, to set open their Cazements into a most delicate Garden and Orchard, wherby they may not only see that, wherin they are so much delighted, but also to give fresh, sweet, and pleasant ayre to their Galleries and Chambers" (56). Such rooms, because of their great length, permitted and were often designed specifically for recreational walking inside the house in poor weather (Mercer 101). The gallery was thus associated with the garden in several ways: by overlooking it, by partaking of its sunlight and breezes, and by allowing pleasant, healthful walks otherwise taken outside, especially in the alleys around the garden that were also called galleries (Amherst 83; Burton and Kelly, *Tudors* 242–3). It is almost as if the garden and its associated virtues were replicated indoors in such rooms. Books identified with the garden might have fit into these spaces quite naturally.

That this was the case may be seen from the fact that Michel de Montaigne once went so far as to dream of installing a gallery within his library, on the grounds that "[e]very place of retirement requires a Walk." Montaigne continues: "My Thoughts sleep if I sit still; my Fancy does not go by it self, [so much] as when my Legs move it" (68). Study and reading, then, were far from denoting bodily confinement in Montaigne's experience; similarly, in a scene from Thomas Middleton's *The Puritan* (1607), one of a group admitted to the gallery of a genteel house exclaims upon "what a fine Gallery were here for mee to walke and study, and make verses" (E3r). Reading itself was often represented metaphorically in terms of the reader's physical motion, which in turn was often characterized as meandering through a flower-filled garden. For instance, in his influential *De copia*, Erasmus conventionally describes a student who "flies about like a diligent bee through the whole garden of authors ... where he would fall on every little flower, collecting a bit of nectar from each to carry to his hive" (qtd. in Bushnell, *Culture* 135–6). John Bodenham expands on the image of the ambulatory reader when he writes "To the Reader" of his *Belvedere* (1600) that

> here thou art brought into the Muses Garden .... The walkes, alleys, and passages in this Garden are almost infinite; every where a turning, on all sides such windings in and out .... Marke then, what varietie of flowres grow all along as thou goest, and trample on none rudely, for all are right precious. (A3^r)

Since just these sorts of "walkes, alleys, and passages"—present in actual as well as in metaphorical gardens—were reproduced indoors in purpose-built galleries, we can infer that such galleries were considered especially appropriate sites for the reading of books like Bodenham's.

Montaigne adds a revealing coda to his previous thought (that his thoughts need his legs to move them) when he speculates that "all those who study without a Book are in the same Condition" (68). That is, in his view, to study without a book is akin to studying books in stillness. The imaginative space created by the book, then, could be imagined as an artificial gallery for the mind to roam in. Since the gallery was so closely associated with and sometimes even a part of the garden, we need only complete the syllogism to see how the book itself could be imagined as a garden laid out for reflective rambling. All this suggests that reading, far from being associated with physical stillness, was or could be associated in the period with healthful exertion. Just as we must work to associate poetry on the page from the period with what we know of its frequent recitation aloud before a group, rather than with our own habits of silent and solitary reading, we may be equally challenged to rethink our image of the stationary reader by recovering historically-specific connotations of the gardens and galleries that are so casually embedded in titles of the period.

If walks taken in the garden could also be taken in the gallery, so could sunlight. Glazed windows had been a luxury in England in the fifteenth century (Beard 17), but "[a]ll houses of any pretensions had glazed windows by 1600" (Peter Thornton 82). Later Elizabethan houses in particular were noted for their capacity to "let in air and sunshine through great expanses of glass" (Beard 36). The notable increase in the size and number of windows would affect the interior domestic atmosphere in several ways, not least by improving the quality of the air circulating in the house, as was noted by several contemporary commentators (such as Lawson, Markham, and Parkinson, quoted above). But the increase of indoor sunlight is perhaps the more significant change with respect to reading. The garden had long been treated as an ideal place for both writing and reading, from a practical standpoint no doubt because of its concentration of natural light.[6]

[6] In the 1594 English translation of Justus Lipsius's *De Constantia*, for instance, the author's interlocutor notes that "the true end and vse of gardens" is "quietnes, with

Yet in most institutions, reading had to occur indoors, where books were often closely chained to desks. In images of such libraries, lecterns tend to be set right against the window, confirming the extent to which books, like plants, craved sunlight.[7] It can be hard to appreciate today how much steady light is needed in order to read—and how little may be had without electricity or the sun. Natural light was, for early modern readers, an imperative but relatively scarce resource; the alternatives were inadequate and sometimes even dangerous to the books they were supposed to illuminate. Lamps and candles flickered, smoked, and dripped; one early modern treatise on libraries expressed despair about the presence of candles near books: "What may I say about people who, overcome by drowsiness while reading, have burned a book to the center of the page!" (qtd. in Perry 132). A sunny place was a safer place to read.

This was perceived to be true for readers as much as for books, since candlelight was thought to be harmful to the eyes of a reader. So much may be inferred from a textbook description of the typical appurtenances of a seventeenth-century study, in which a scholarly inhabitant, "[b]eing to sit up late … setteth a Candle, on a Candle-stick" and "before the Candle he placeth a Screen, which is green, that it may not hurt his eyesight" (Comenius, *Joh. Amos Commenii Orbis Sensualium Pictus* 201). Similarly, the interlocutor of Justus Lipsius notes in his texts that, "[a]s painters hauing dimmed their eies with long and earnest beholding their work, do recomfort them with certain glasses or green colours[,] so here"—the setting being a lush and blooming garden—"may we refresh our wearied and wandring minds" (66). Like the accountant's green visor or the banker's green shaded lamp of later eras, a green light in the early modern period, even one refracted from botanical surroundings, was thought to ease the task of visually processing text. That the garden's ability to refresh the eyes was one reason why it was considered an optimal space to read may be further inferred from the words of the seventeenth-century herbalist William Coles, who suggested that

> if a man be wearied with over much study ... there is no better place in the world to recreate himselfe than a Garden .... If his sight be obfuscated and dull, as it may easily be, with continuall poring, there is no better way to relieve it, then to view the pleasant greenesse of Herbes, which is the way that Painters use, when they have almost spent their sight by their most earnest contemplation ... (115)

It could thus be a clever rhetorical move for a book, the source of similar eye-strain, to construe itself instead as the site of its relief by characterizing itself as a garden or grove. This aspect of the metaphor, and of early modern experiences

---

drawing from the world, meditation, reading, and writing": "The studious and learned wits of our age ... delight in gardens; and in them (for the most part) are compiled those diuine writings of theirs .... So many sharp and subtil disputations of naturall philosophy, proceed from those greene bowers .... Here you learned Poets compose yee some poemes worthy of immortalitie" (65–6).

7 See examples in Petroski (118) and Peter Thornton (269).

of reading, may lend Marvell's "green thought in a green shade" (48) an entirely new context and meaning.

Even as dedicated a reader as Montaigne, who had "[l]ight very finely fitted" in his library so that he could "there pass away both the most of the days of [his] Life, and most of the hours of those days," was unable to continue reading when the sun went down; he adds a perfunctory and melancholic note to the previous passage: "In the Night I am never there" (68). Books may be uniquely like plants among household objects in their special dependence on sunlight. One advantage of private book ownership for the average early modern reader, unequipped with the likes of Montaigne's towering and artificially illuminated library, would have been the ability to read a book in the optimal conditions of his own backyard or the gallery that mimicked the garden's virtues indoors. Alternatively, he might prefer to read in the luxuriously private conditions of his own bedroom, which was often placed near the gallery (Cooper 302). That books were commonly read in bed may be inferred from a sample conversation included in the language textbook entitled *Florios seconde frutes, to be gathered of twelue trees, of diuers but delightsome tastes to the tongues of Italians and Englishmen. To which is annexed his Gardine of recreation yeelding six thousand Italian prouerbs*:

> L. Doo you want anie thing? Shal I put out the candle?
> M. No truely, but let the candle alone, for I will reade a Chapter.
> L. What booke will you reade now you are a bed?
> M. The Bible, I can not fall asleepe without reading. (161)

Since such textbooks are designed to provide their readers with useful phrases, one can only conclude that to read in bed was quite an ordinary activity. Elizabeth Hardwick, Countess of Shrewsbury, is known to have kept her few books in her bedroom (Peter Thornton 396). Such a room might be known, moreover, as either the solar (Beard 6) or paradise (Cooper 301, 345).[8] The latter name derives, apparently "from its location at the top of the house" (Cooper 301), but perhaps both terms refer to the unusual amount of light that reached the room. Books with titles like *The paradyse of daynty deuises* (Edwards) and *A paradise of praiers and meditations* (Maihew) may thus have been advertising their suitability for bedside reading.

Stillroom and Study

There were countless period uses for plants in the domestic interior, from rushes cut for lights to thistledown gathered for pillows (Thomas 73). An allusion to plants would thus have been at least as likely to evoke practical matters as the

[8] In the sixth definition for "Paradise" given by the *OED*, however, an example by John Leland from 1538 is given that identifies not a bedroom but a study as such: "I saw in a litle studiyng Chaumber ther caullid Paradice the Genealogie of the Percys."

more aesthetic experiences discussed above. Overarching the miscellaneous everyday uses of plants was the vital fact that they sustained life and health from harvest to harvest; this was, after all, "a time when most remedies were manufactured at home in the stillroom" (Arber, *Herbals* 39). The preserving of fruits and distilling of herbs were sufficiently important and involved activities to require a separate, specialized room at a time when few household activities were granted such a privilege (Amherst 107); at least "every manor house from the fifteenth century onward boasted of a still-room where herbal preparations were made and stored for future use" (Frank Anderson 119). The stillroom reveals another meaningful nexus between plants and books and the place of both in the early modern domestic interior.

The activities surrounding the gathering and preserving of plants would have been urgent: many plant products for food and medicine required harvesting precisely when they were in season, since they would not return for another year, if then, and methodical storage was required for the nutritious and curative virtues of the plants to be preserved. Before artificial refrigeration and vacuum-sealing, the preservation of organic materials was a complex and skilled affair, one reflected in "a perfect wave of preservation recipes—for pickling, potting, and spicing—[that] washes over the cookery of the sixteenth century" (Hartley 312; see also Cook, "Time's Bodies"). The gnawed, pitted, rotting fate of books that were poorly stored was known to be similar to that of plant materials that were similarly treated. It may be, then, that another allusion embedded in the metaphor of the garden of verse would have been the necessity of treating books with the same care as was given to the preservation of valuable plants and botanical ingredients. Such preservation was of special concern owing to the tendency for much printed material to be disposed of casually as wrapping paper or for blocking drafts. These fates had become especially visible around the middle of the century in the wake of the ecclesiastical dissolution, when "tens of thousands of manuscripts ... perished" after being used "to polish candlesticks, and boots, to wrap pies, to press gloves flat, or to repair broken windows" (Baker 196). At this time, too, "wallpaper, as well as box linings and the inside of book bindings, were favoured as ways of using sheets from books that ran counter to established political and religious opinions" (Rota 6; see also Cressy 94). Far from signaling a need for careful preservation, on the other hand, the metaphor of the botanical book might have amounted to a wry acknowledgment of the ephemeral nature of what was printed. Everard Guilpin's seventieth epigram in his *Skialetheia* (1598) reflects his awareness of the likely fate of some of his verses and shows another close connection between poetry and plants: "thow'lt doome them to th'Apotheta"—likely a misprint or variant of Apotheca, Latin for apothecary's shop—"To wrap Sope in, and Assifœtida" (5–6), the latter being a medicinal spice; he adds that he knows very well that his "vnsauory rimes: / [are] Fit to wrap playsters, and odd vnguents in" (10–11). Pages from books of verse such as his may thus have been employed and encountered as wrappings on herbal medicines, such that these pages would contain, in different ways, both medicines and texts. This fact may go some way to explaining the tendency in the period to imbue texts with pharmaceutical powers, which might quite literally have been thought to "rub off" on them.[9] It may also set in a new context the tendency in the period to liken a library to a pharmacy, as in Claude Clement's early seventeenth-century treatise on library theory, which advises the inscription above a library's entry of mottoes including "Pharmacy of the Soul" (Rovelstad 546).

A stillroom's shelves would have carefully organized and labelled contents, much like those of a library, and it would almost certainly include at least one book. A study by Peter Clark of the location of books in the homes of Elizabethan townsfolk uncovered that not a few were kept in the kitchen, another room associated with the processing of vegetable materials, and that their number in this location was on the rise (103–4; see also Siraisi 148). These books would no doubt include the household herbal, or more than one of its kind as cheap printed versions became increasingly available. The older, hand-written sort of herbal or cookbook was often integrated with other household uses, and therefore had a markedly anthological quality: recipes and cures, along with scraps of verse and jokes, were added and altered over the years by various hands. Roy Porter notes that "[a]t least from the sixteenth century, it was common for gentry families to keep personal manuscript recipe books, where, cheek by jowl with cooking tips ('receipts') and household hints, there could be found medical remedies, many of them culled from learned sources" (2). Such books could also, moreover, contain verse: Marotti refers to manuscript poetical anthologies including "catch-all miscellanies containing such other material as prose letters, medical recipes, household accounts, school exercises, and journal writing" ("Malleable" 162). Earle Havens notes that "medicinal and culinary recipes" are frequently intermingled among the entries in commonplace books from this period, such that between fragments of pleasurable and profitable verse one might happen upon a herbal cure (9). Similarly, in 1620, *A Helpe to Memory and Discourse* was published; the extended title describes the text as having been "extracted from the sweating braines of phisitions, philosophers, orators and poets: distilled in their assiduous and witty collections" (Basse). The authority of the words of physicians and poets are here set on par, as they are in the herbal of John Gerard (see Chapter 4); moreover, the words of both are imagined to be subject, like herbal simples, to a metaphorically pharmaceutical process of distillation.

There may have been in the period an understood affiliation, both architectural and functional, between the stillroom and the study. In 1534, for instance, Cardinal Fisher had three rooms called studies, but the inventory indicates that one was a medical storeroom (Cooper 300). Stillrooms were also sometimes called "closets," which were also common as sites for reading and the keeping of books. Alan Stewart, in his exploration of the uses of sixteenth-century English closets, cites

---

[9] Books were, at this time, "commonly treated as possessing intrinsic healing capacities" (Porter 9), as may be seen in some titles, such as Henry Holland's *Spiritual Preservatives against the Pestilence* (1593) and Roger Fenton's *A Perfume Against the Noisome Pestilence* (1603), as well as some of William Turner's titles noted in Chapter 3.

the 1556 inventory of the "his and hers" closets at Loseley Hall: that of Sir William More's wife contains, among other things, "glasses, pots, bottles, jugs, conserve jars, sweetmeat barrels, an hourglass, a grater, knives, a pastry-mould"—objects suggestive of the work of canning and preserving—along with five books including "the pomeaunder of prayers," a book by Thomas Becon with a title that suggests that it is the product of just such plant-processing activities (82). Similarly, a century later, *The closet of the eminently learned Sir Kenelme Digbie Kt. opened* advertises itself on its title page as having *discovered several ways for making of metheglin, sider, cherry-wine, &c.: together with excellent directions for cookery, as also for preserving, conserving, candying, &c*; despite his wide-ranging studies and diplomatic activities, Digby's "closet" is imagined as primarily a place not for reading so much as for working with plant ingredients.

The stillroom's hoard of botanical medicines and the books associated with them epitomized impeccable necessity. Together, such books and botanical remedies represented and assured household readiness against the dangers and deprivations of illness and winter, two conditions that entail extraordinary confinement to the domestic interior and that therefore make such diversions as books of verse and such comforts as devotional books seem as indispensable as any herbal book or medicine. Yet because many herbal preparations were often little more than liquors or sweets, even apparently rigorous regimens could also yield pleasure. Books that claimed to offer a similar mix of pleasure and therapy—like Roger Fenton's *Perfume Against the Noisome Pestilence*—might find it that much easier to find shelf-space in households that were already prepared to accommodate an arsenal of herbal preparations. For a book to be likened to plants in this context was to claim to be useful, even essential, on a daily or at least occasional basis, and thus best kept on hand against future need. However, as Sharon Achinstein has detailed, some early modern texts could also be treated, conversely, as agents of disease: "Renaissance notions of contagion and transmission linked plagues to ballads; the evil in plagues and ballads was thought to disseminate in similar ways" (27). There was, moreover, not just a metaphorical or moral but a material association made between books and the deadly disease: "the very physical object itself was a potential carrier of disease, since the rags from which paper was made were thought to be able to carry the plague" (34). Achinstein also notes the extent to which the prevention of plague entailed the prevention of its entry into the domestic realm: "The quarantines and orders for segregation required that infected persons be shut up inside their own houses, the healthy with the sick, and a lock put on the outside door .... The plague was domestic; it attacked the house" (33). In a period when fatal outbreaks occurred almost every year (27), books that portrayed themselves as physic against such disease might have had a more specific domestic appeal and applicability than is otherwise apparent; they might also have stood a better chance of escaping castigation, or burning, as bearers of the disease they claimed to cure.

The advice of William Cole, cited above, to proceed for the sake of health from a place of study to the garden suggests the perceived advantage of a close proximity between domestic spaces devoted to these things: the latter provided salubrious respite from the former. There appears to be a similar tendency in many images and texts of the period to represent gardens and studies as contiguous spaces, ideally suited to be placed adjacent to one another. Montaigne himself, referring to his tower library, wrote that "I thence under me see my garden" (68). Francis Bacon has one of his counsellors to the king link an imaginary library and garden as ideally complementary spaces in his *Gesta Grayorum*, first published in 1594:

> First, the collecting of a most perfect and general library, wherein whosoever the wit of man hath heretofore committed to books of worth … may be made contributory to your wisdom. Next, a spacious, wonderful garden, wherein whatsoever plant the sun of divers climate, or the earth out of divers moulds, either wild or by the culture of man brought forth, may be ... set and cherished. (34–5)

Another imaginary continental integration of garden and library can be found in the ideal garden, planned in the mid-sixteenth century by Bernard Palissy, in which "would be seen, engraved, carved or formed from cut branches, texts taken from the two sapient books—Proverbs and Ecclesiastes" (Lecoq 69); Palissy went so far as to imagine integrating the very physical forms of plants and texts. One part of the plan involved a cliff with hollowed chambers on two levels, the lower acting as orangery and granary, places to grow and store plant material, the upper as "bookshop and study, shops selling spirits, vinegars, dried fruits and so on, and a laboratory for distilling herbs" (Lecoq 71). Once again, the activities of reading were conjoined with the activities of processing plants. In this garden, as in his better-known ceramics, Palissy attempted to fuse nature and art, this time in the form of plants and texts.

In the Villa Brenzone at Punta San Vigilio could be found a similar link between the garden and the library. Part of the grounds of the estate of the philosopher Agostino Brenzone built in the 1550s consisted of a plateau encircled by trees and a wall fitted with antique busts and inscribed plaques, a space that "evokes a small reading room, here open to the sky, but nevertheless based on the small chamber with circular plan and domed ceiling to be found in the palaces of the Veneto" (Visentini 107). The ideal form of the sixteenth-century study and that of the garden coincided. Similarly, in her study of the study in Renaissance Italy, Dora Thornton notes the "close links between the study and gardens, particularly in the context of the villa" (30). Thornton confirms Cole's suggestion that it was not just the two architectural spaces but the activities they facilitated that were seen to be ideally linked: "Scholars liked to turn from their studies to the sensuous delight and relaxation of a garden" (31). As an example, Thornton quotes Angelo Decembrio's mid-sixteenth century description of a courtier's ideal villa, in which "there is daily conversation about books, and in this house there is a library—as you see, looking out over the greenest little plot" (31). A visit is portrayed as proceeding from the picking of fruit in the garden to an exploration of the library, which has itself been botanically camouflaged: "Waste soon having been laid to the fruit, we went up into Giovanni's library, whose floor and books he himself had strewn with the white and mauve flowers we call 'chariophilos'" which emit "a wonderful fragrance smelling as if of grace and love" (translated in Perry 142). These plants are not an incidental but a functional part of such libraries; Decembrio offers "detailed recommendations for the use of plant materials in the library to improve the atmosphere and protect the books" (121). One of the speakers, for instance, advises the use of "white lilies, and cypress panels ... cuttings of rosemary and myrtle too, and bunches of roses or violets, and sweet fruits, especially ones referred to as citrus" to combat the bad smells that accumulate in libraries and "wormwood, rue, and bay laurel and its berries" to defend against insect predators (131). Perry summarizes the atmospheric effect of such a library: "the garden has been brought indoors" (160).

Comparable evidence for the integration of plants and books in the life of English readers is harder to find. Perry notes, however, that "the association of plants and plant products with libraries and the preservation of books is not unknown" in sixteenth-century England, since those who kept Queen Elizabeth's books were "referred to in official documents as 'distillers of odiferous herbs'" (Perry 138); I have been unable to find more detail on this suggestive connection. One contemporary English example of a study integrated with horticultural functions is, however, just as illuminating in its own way: this is one particular "Studdy Roome" inventoried in 1610 and found to contain two desks, including one "wereuppon the Book of Martyrs now standes," along with two "fayre tronckes," likely also designed to hold books, including perhaps some of the twenty-four itemized in the inventory; one who has studied the inventory notes that "[w]ith all this evidence of learning"—including an ostrich egg—"... it comes as something of a surprise to find that there was also 'a black pyk for hay ... behind the Study-room dore'" (Peter Thornton 303). But given the widespread association between study and stillroom, the sharing of space by this horticultural implement and the tools of book-learning was very likely less of a surprise at the time.

Posies Transposed

A popular Tudor ornament bridging the gap between the botanical and the textual—one found everywhere from walls to painted wooden dishes—was the posy. This was not a bouquet but a sententious saying, often in verse and usually inscribed on some material other than a page in a book (Byrne 46; Fleming, "Wounded" 6). Nicholas Bacon, for example, famously had his long gallery at Gorhambury inscribed with them (McCutcheon). It has been suggested that such textual inscriptions "prepared the way for hanging galleries with pictures" (Cooper 320); they thus spanned the gap between text and image by embodying writing in unexpected material forms. In his *De Ratione*, Erasmus advised them as memory aids for the student, who "should inscribe apophthegms, proverbs, and sententiae at the front and back of books, and engrave them on rings and cups, and paint them [on] doors and walls and even on window glass, so that nowhere could one avoid what would add to learning" (qtd. in Bushnell, *Culture* 133–4). Expanding their botanical association beyond the basic floral pun, posies also had a special association with the consumption of fruit: George Puttenham, in *The Arte of English Poesie*, refers to posies being "printed upon ... banketting dishes of suger plate" or more often painted "upon the backsides of our fruit trenchers of wood" (72). In his verse letter to Henry Goodyere, John Donne offers his friend advice, which he acknowledges as commonplace when he suggests that "Tables, or fruit-trenchers teach as much" (163). Thomas Tusser's *A hundreth good points of husbandry lately maried vnto a hundreth good poynts of huswifery* (London 1570) ends with a compilation of "Husbandly posies" divided up according to the domestic space they are designed to inhabit: "posies for the hall," "Posies for the Parler," "Posies for the geastes chamber" and "Posies for thine own bed Chamber" (39–40). Despite the destruction of most houses from this period, considerable evidence remains of this advice being taken up.[10]

Watt dates the heyday of domestic textual inscription between 1560 and 1580, following its widespread imposition in churches in the wake of Reformation iconoclasm. She suggests that "[t]he acknowledged purpose" of such texts

> was not only didactic, but aesthetic, to fill the whitewashed spaces under which the idolatrous wall paintings now lay. In the absence of pictures, words themselves could take over a decorative function .... The new text-based aesthetic of the church was transferred to domestic interiors. (217–18)

The ecclesiastical precedent for textualizing one's surroundings would have given considerable status to the situating of texts in homes. As well as painting texts on walls, the early modern English attached pages themselves to walls (Watt 221–3). Anne Clifford went still further, according to her eulogist:

> [s]he would frequently bring out of the rich Store-house of her Memory ... Sentences, or Sayings of remark, which she had read or learned out of Authors, and with these her Walls, her Bed, her Hangings, and Furniture must be adorned; causing her Servants to write them in Papers, and her Maids to pin them up, that she, or they, in time of their dressing, or as occasion served, might remember, and make their descants upon them. So that, though she had not many Books in her Chamber, yet it was dressed up with the flowers of a library. (Rainbowe 40)

While Clifford's posies were manuscripts, Watt has detailed the concurrent use of printed sheets of paper as wall hangings (Chapter 5). The posy became visible in various aspects of Elizabethan domestic interiors of all social ranks (Watt 148). In posy form, the written word was habitually displayed in household interiors, outside of books and before books themselves were as commonly displayed or owned. One consequence of this widespread presence of textual ornament was that "fragments of text became furniture—the cluttered movables of the ordinary house and the educated mind" (Bushnell, *Culture* 134). Because a "posy" equally

[10] See Juliet Fleming's *Graffiti and the Writing Arts of Early Modern England*, especially Chapters 1 and 2.

signified a household verse inscription and a gathering of flowers, as well as a homonym for poetry in general, it was at the heart of the period's association of interiors, plants, and texts. This type of text, identified by means of a pun that linked verse and floral gatherings, provided a domesticated precedent for books that identified themselves in related botanical terms.

The early modern convention of the garden of verse can thus be read as part of a broader convention of depicting the book as a room: as treasury or closet, cabinet or storehouse, and later, of course, as magazine. The form of the book itself has long been imagined as an interior—and, as these terms reveal, often as a domestic one. Many printed title-pages of the period were framed by architectural imagery, so that the opening of the book paralleled opening a door (Shevlin 52; Sherman). The analogy between books and architectural interiors had a particularly physical basis in the early modern era, when many bound books were emphatically enclosed spaces, with heavy wooden covers and intricate clasps. A historian of book furniture has established that "until the end of the sixteenth century, fastenings were commonplace on ordinary wooden board bindings" (Durrfeld 305); the book, like the typically enclosed garden of the period, had to have its gate unlatched before it could be entered. A visual analogy exists between the arrangement of such book furnishings and the lay-out of some Tudor garden beds. A typical binding is described in this way: "The back board of the folio binding is mounted with four square corner pieces and one oval central piece .... The front board of the octavo binding, closed with one all-metal fastening attached to its fore-edge, is mounted with four round bosses in the corners and one large central piece" (Durrfeld 311). Such an arrangement echoes one common garden design of the period, with a circular central area (often a fountain) surrounded in four corners by geometrically-contoured beds. Even the terminology used in discussing architecture and textuality overlapped: construction work was known as "framing" (Beal 5), a key term in the humanist handling of texts (Crane, *passim*).

One materially-minded historian of the early modern writing arts has suggested that "Elizabethans ... saw most or all writing as having both dimension and location" (Fleming, "Wounded Walls" 2). Another scholar of Renaissance garden history has suggested, similarly, that not just writing but the entire "mental world [was] conceived as a spatial structure, a network of compartments or places," such that "to understand any given matter [was] to remember its position, to see its natural place in the order of things" (Comito, *Idea* 71). In late sixteenth-century England, however, newly affordable, plentiful, and popular books lacked a "natural place" in the household—unlike posies painted on walls or those gathered from gardens. One way book-makers might have promoted their wares in such an environment was to define a place for their products by associating them with something more firmly entrenched in domestic life. The garden and its produce had long occupied a spectrum of delightful and profitable spaces in the home; and books, in the course of the period, came to do the same.

Not long after 1600, gardens appear to have lost some of their cachet as metaphors for books, while other metaphors took their place. In William H. Sherman's study of the way in which early modern paratextual materials "were understood in spatial—and often specifically architectural—terms" (72), he finds not gardens but "a veritable vogue for textual porches" in the mid-seventeenth century (77). Similarly, Alan Stewart discusses another domestic space gaining prominence over gardens in the miscellanies of the early seventeenth century: "now readers were tempted by 'closets' 'cabinets' [and] 'studies'" (89). In both cases, the metaphor shifts from the garden to a place closer to or more intimately a part of the domestic interior. While the porch may be seen, like the garden, as a kind of "transitional zone between inside and outside" (Sherman 77), the closet or cabinet was in many ways the most intimately interior of all domestic spaces (Cooper 300). These shifts in metaphor may signal a larger shift in the degree of cultural comfort with the domestication of the book in early modern England.

Epilogue

With the spread of print in the sixteenth century, early modern English men and women were increasingly exposed both to books and texts about plants and to books and texts that construed themselves, metaphorically, in terms of plant material. A better understanding of the bookishness of early modern botanical culture helps to clarify the historically-specific connotations of the rhetoric of plant collection that permeates so many non-botanical texts from the same period. I have placed special emphasis in the preceding chapters on the way in which the historical and emerging nature of sixteenth-century medicinal herbalism meant that plants in the period always hit the ground already highly mediated by texts. This conventional botanical bookishness was both highlighted and reinforced by the innovative genre of the herbarium in which plants were fused to the very material of the book, while in the printed herbals of the period, classical and medieval literary traditions about plants were preserved, reproduced, and reformed by the writing practices of herbalists. While William Turner has been hailed as the originating author of a newly scientific botany in England, John Gerard reveals the continued strength of the herbal literary tradition half a century later in his herbal's ingrained poetic elements and its strong resemblance, in its composition, to the commonplace book and the anthology—two types of book commonly understood to be composed of textual and rhetorical "flowers." The increased prominence of and value placed on plants in domestic collections helps to explain why so many of the increasing quantity of books printed in the period liken themselves to gatherings of plants in order to define their role and value as cultural artifacts.

The simple fact that the two disparate materials of plant and book were so frequently compared in this period is not, in itself, surprising; as Foucault long ago detailed, the sixteenth-century European *episteme* "consisted of the reciprocal cross-reference of signs and similitudes" (33). Foucault's fruitful claim suggests that the cultural historiography of both plants and books may be enriched by considering them, as they were considered in the early modern period, in light of one another. The early modern interpenetration of these two types of object was so thorough as to extend beyond the metaphorical to the material level: in the composition of inks, pages, and covers from plant materials, in the innovative fusion of plant materials in *horti sicci*, and in the parallel and sometimes overlapping placement of plants and books in the home. At a time when a culturally-dominant metaphorical mindset meant that "books were never just books" and "print was not just print" (Watt 251), the physical resemblances and interpenetrations of books and plants would have made them appear especially apposite materials for comparison. An understanding of the variety, concreteness, and prevalence of the intersections between early modern plants and books helps to account for the unusual concentration of conceptual traffic between these two otherwise dissimilar objects.

Many sixteenth-century intersections between botanical and bookish discourses would be severed in the seventeenth century. The resemblance between artistic and scientific disciplines and their amalgam of fact and fiction proved the focal point for extensive critique by later students of the same disciplines as they were reconceived when a theoretically text-repudiating empiricism tried to erase its humanist and literary ancestry. The presence of poetic material in herbals would be criticized and eventually eliminated by those following in footsteps of Francis Bacon (Shapiro 108–11, 160–67). Connections between words and things would be subject to strenuous new analyses in which the verbal element was judged to be of best service when subordinated or replaced with mathematics. Ancient authors were to be less authoritative than contemporary eye-witnessed experience, while evocative figurative descriptions perceived by the naked eye were to be supplemented and sometimes supplanted by objective measurements taken under the microscope and by a technical terminology corresponding to an analysis, not of whole plants anymore, but of increasingly minute parts of them. Makers of cutting-edge herbaria eventually cut the threads that bound their pages to a spine, disposing of the formal bounds of the codex.

Subsequently, and in many cases consequently, Western cultural understandings and uses of plants were transformed. As plant nomenclature grew increasingly rigorous in the eighteenth, nineteenth, and twentieth centuries, plant study gradually became less of a popular cultural influence and the practical medical element of botany was eventually almost eclipsed. Broader human relations to plants have, in the industrialized world, mainly been obliterated or obscured in daily life, since individuals rarely process plants as a primary resource in an age when carrots are bought already plucked and peeled, bite-sized and bagged. At the same time as personal interaction with the plant world has become so much more mediated and attenuated than it was five hundred years ago, the written word has secured its cultural dominance, not least through increasingly productive technologies that outpace the once outstanding reproductive capacity of plants themselves. In the wake of what may be a reversal of the cultural priority of these two materials, it may be difficult to appreciate the extent to which the bookish could be conceived of in terms of the botanical in the early modern period. Consequently, perhaps, textual collections and botanical collections have generally been studied as separate cultural fields and practices. I hope that this study goes some way towards overcoming the separation by clarifying the symbiotic nature of plant culture and print culture in early modern England.

Bibliography

Achinstein, Sharon. "Plagues and Publication: Ballads and the Representation of Disease in the English Renaissance." *Criticism* 34.1 (1992): 27–49.

Adams, Thomas. *A Divine Herball*. London, 1616.

Addison, Joseph and Richard Steele. *Addison and Steele: Selections from* The Tatler *and* The Spectator. Ed. Robert J. Allen. 2nd ed. New York: Holt, Rinehart and Winston, 1970.

Alexander, Catherine. "The Garden as Occasional Domestic Space." *Signs: Journal of Women in Culture and Society* 27.3 (2002): 857–71.

Amherst, Alicia. *A History of Gardening in England*. London: Bernard Quaritch, 1896.

Amtower, Laurel. *Engaging Words: The Culture of Reading in the Later Middle Ages*. New York: Palgrave, 2000.

Anderson, Frank J. *An Illustrated History of the Herbals*. New York: Columbia UP, 1977.

Anderson, Randall L. "Metaphors of the Book as Garden in the English Renaissance." *Yearbook of English Studies* 33 (2003): 248–61.

Anon. *A godlie gardeine*. London, 1569.

Anon. *The grete herball*. London, 1526.

Appleby, Andrew B. "Diet in Sixteenth-Century England: Sources, Problems, Possibilities." Ed. Charles Webster. *Health, Medicine, and Mortality in the Sixteenth Century*. Cambridge: Cambridge UP, 1979. 98–116.

Arber, Agnes. "From Medieval Herbalism to the Birth of Modern Botany." Rpt. in *Herbals: Their Origin and Evolution*. Cambridge: Cambridge UP, 1953. 317–36.

———. *Herbals: Their Origin and Evolution: A Chapter in the History of Botany, 1470–1670*. 3rd ed. Cambridge: Cambridge UP, 1986.

Ashworth, William B., Jr. "Emblematic Natural History of the Renaissance." *Cultures of Natural History*. Ed. Nick Jardine, James Secord, and Emma Spary. Cambridge: Cambridge UP, 1996. 17–37.

———. "Natural History and the Emblematic World View." *Reappraisals of the Scientific Revolution*. Ed. David C. Lindberg and Robert S. Westman. Cambridge: Cambridge UP, 1990. 303–32.

Atran, Scott. *Cognitive Founations of Natural History: Towards an Anthropology of Science*. Cambridge: Cambridge UP, 1990.

Austen, Ralph. *The Spirituall Use of an Orchard, or Garden of Fruit-Trees*. London, 1657.

Avrin, Leila. *Scribes, Script and Books: The Book Arts from Antiquity to the Renaissance*. Chicago: American Library Association, 1991.

A.W. *A book of cookrye*. London, 1591.

Bacon, Francis. *The Advancement of Learning*. London, 1605.

———. *Gesta Grayorum*. 1594. Ed. William Canning. London, 1688.

———. *A Selection of His Works*. Ed. Sidney Warhaft. Toronto: Macmillan, 1965.

———. *Sylva Sylvarum, or, A Naturall Historie in Ten Centuries*. London, 1627.

Baker, Nicholson. *The Size of Thoughts*. London: Chatto and Windus, 1996.

[Banckes' herbal.] *Here begynnyth a newe mater, the whiche sheweth and treateth of ye vertues [and] proprytes of herbes, the whiche is called an herball.* London, 1525.

Barker, William. *The VIII. Bookes of Xenophon*. London, 1567. N. pag.

Barlow, H.M. *Old English Herbals, 1525–1640.* London: Bale, 1913.

Basse, William. *A Helpe to memory and discourse*. London, 1620.

Beal, Peter. "Notions in Garrison: The Seventeenth-Century Commonplace Book." *New Ways of Looking at Old Texts: Papers of the Renaissance English Text Society, 1985–1991*. Ed. W. Speed Hill. Binghamton, NY: Renaissance English Text Society, 1993. 131–47.

Beard, Geoffrey. *The National Trust Book of the English House Interior*. London: Viking, 1990.

Becon, Thomas. *Flour of Godly Praiers*. London, 1550. N. pag.

Belk, Russell W. *Collecting in a Consumer Society*. London: Routledge, 1995.

Benedict, Barbara M. *Curiosity: A Cultural History of Early Modern Inquiry*. Chicago: U of Chicago P, 2001.

———. "Literary Miscellanies: The Cultural Mediation of Fragmented Feeling." *ELH* 57 (1990): 407–30.

———. *Making the Modern Reader: Cultural Mediation in Early Modern Literary Anthologies*. Princeton: Princeton UP, 1996.

Bennett, H.S. *English Books and Readers, 1475 to 1557*. Cambridge: Cambridge UP, 1969.

Bennett, Jim and Scott Mandelbrote. *The Garden, The Ark, The Tower, The Temple: Biblical Metaphors of Knowledge in Early Modern Europe*. Oxford: Museum of the History of Science in Association with the Bodleian Library, 1998.

Beretta, Ilva. '*The World's a Garden': Garden Poetry of the English Renaissance*. Acta Universitatis Upsaliensis, Studia Anglistica Upsaliensia 84. Uppsala: Uppsala University, 1993.

Birrell, T.A. "Reading as Pastime: The Place of Light Literature in Some Gentlemen's Libraries of the 17th Century." *Property of a Gentleman: The Formation, Organisation and Dispersal of the Private Library 1620–1920*. Ed. Robin Myers and Michael Harris. Winchester: St. Paul's Bibliographies, 1991. 113–31.

Bishop, John. *Beautiful blossomes*. London, 1577. N. pag.

Blacker, Mary Rose. *Flora Domestica: A History of British Flower Arranging*. London: National Trust, 2000.

Blair, Ann. "Annotating and Indexing Natural Philosophy." *Books and the Sciences in History*. Ed. Marina Frasca-Spada and Nick Jardine. Cambridge: Cambridge UP, 2000. 69–89.

Blair, Ann. "Humanist Methods in Natural Philosophy: The Commonplace Book." *Journal of the History of Ideas* 53 (1992): 541–51.

Blank, Paula. *Broken English: Dialects and the Politics of Language in Renaissance Writings*. London: Routledge, 1996.

Blayney, Peter. *The Bookshops in Paul's Cross Churchyard*. London: Bibliographical Society, 1990.

Blunt, Wilfrid. *The Art of Botanical Illustration*. London: Collins, 1950.

Blunt, Wilfrid and Sandra Raphael. *The Illustrated Herbal*. Toronto: Oxford UP, 1979.

Bock, Hieronymus. *Kreuterbuch, darinn Underscheidt, Namen und Würckung de Kreuter.* Strassburg, 1551.

Bodenham, John. *Bel-vedére or, The garden of the muses*. London, 1600.

Boehrer, Bruce Thomas. *The Fury of Men's Gullets: Ben Jonson and the Digestive Canal*. Philadelphia: U of Penn P, 1997.

Boesky, Amy. "'Outlandish-Fruits': Commissioning Nature for the Museum of Man." *ELH* 58 (1991): 305–30.

Bono, James J. *The Word of God and the Languages of Man: Interpreting Nature in Early Modern Science and Medicine*. Madison: U of Wisconsin P, 1995.

Booth, F.E.M. and G.L. Lucas. "The Role of Botanic Gardens in Economic Botany." *Plants and Society*. Ed. M.S. Swaminathan and S.L. Kochar. London: Macmillan, 1989. 452–70.

Boutroue, Marie-Elisabeth and Jacques Gana. *L'Herbier de Paolo Boccone*. [Online exhibition.] Paris: Bibliothèque interuniversitaire de médicine, July 2001. <http://www.bium.univ-paris5.fr/boccone/>

Braudel, Ferdinand. *Capitalism and Material Life, 1400–1800*. Trans. Miriam Cochan. New York: Harper and Row, 1973.

Bremer, Jan Maarten. "From Stony Facts to Paper Flowers." *Arcadia* 38 (2003): 304–13.

Breton, Nicholas. *The arbor of amorous deuises*. London, 1597.

———. *Brittons bowre of delights*. London, 1591.

———. *A smale handfull of fragrant flowers*. London, 1575.

Bright, David F. *Elaborate Disarray: The Nature of Statius' Silvae*. Meisenheim am Glan: A. Hain, 1980.

Brückner, Martin and Kristen Poole. "The Plot Thickens: Surveying Manuals, Drama, and the Materiality of Narrative Form in Early Modern England." *ELH* 69 (2002): 617–48.

Bullein, William. *Bulleins bulwarke*. London, 1562. N. pag.

———. *A newe boke of phisicke called ye gouernment of health*. London, 1559.

Burton, Elizabeth and Felix Kelly. *The Early Tudors at Home, 1485–1558*. London: Allen Lane, 1976.

———. *The Elizabethans at Home*. London: Secker and Warbur, 1958.

Burton, Robert. *The Anatomy of Melancholy*. Ed. Holbrook Jackson. New York: New York Review Books, 2001.

Bushnell, Rebecca. *A Culture of Teaching: Early Modern Humanism in Theory and Practice*. Ithaca: Cornell UP, 1996.

———. *Green Desire: Imagining Early Modern English Gardens*. Ithaca: Cornell UP, 2003.

Byrne, M. *Elizabethan Life in Town and Country*. 1925. 8th ed. London: Methuen, 1961.

Camille, Michael. "Sensations of the Page: Imagining Technologies and Medieval Illuminated Manuscripts." *The Iconic Page in Manuscript, Print, and Digital Culture*. Ed. George Bornstein and Theresa Tinkle. Ann Arbor: U of Michigan P, 1998. 33–47.

Campbell-Culver, Maggie. *The Origin of Plants*. London: Headline, 2001.

Cantor, Leonard. *The Changing English Countryside 1400–1700*. London: Routledge and Kegan Paul, 1987.

Carlson, Eric Josef. "The Marriage of William Turner." *Historical Research* 65 (1992): 336–9.

Carter, John and Percy H. Muir, ed. *Printing and the Mind of Man*. London: Cassell, 1967.

Cartwright, Thomas. *An hospital for the diseased*. London, 1578.

Castillo, Jesús Carrillo. "Naming Difference: The Politics of Naming in Fernandez de Oviedo's *Historia General y Natural de las Indias*." *Science in Context* 16 (2003): 489–504.

Cave, Terence. *The Cornucopian Text: Problems of Writing in the French Renaissance*. 1979. Oxford: Clarendon, 1986.

Chapman, George T.L. and Marilyn N. Tweddle, ed. Introduction. *A New Herball*. By William Turner. Manchester: Mid Northumberland Arts Group and Carcanet, 1989. 8–22.

Charlesworth, Michael, ed. *The English Garden: Literary Sources and Documents*. Vol. 1: Chronological Overview: 1550–1730. Mountfield, East Sussex: Helm Information, 1993.

Chartier, Roger. *The Cultural Uses of Print in Early Modern France*. Trans. Lydia G. Cochrane. Princeton: Princeton UP, 1987.

Cicero. *Rhetorica ad Herennium*. Trans. Harry Caplan. Cambridge, Mass.: Harvard UP, 1954.

Clark, J.W. *The Care of Books: An Essay on the Development of Libraries and their Fittings, from the Earliest Times to the End of the Eighteenth Century*. Cambridge: Cambridge UP, 1901.

———. *Libraries in the Medieval and Renaissance Periods*. Chicago: Argonaut, 1968.

Clark, Peter. "The Ownership of Books in England, 1560–1650: The Example of Some Kentish Townsfolk." *Schooling and Society: Studies in the History of Education*. Ed. Lawrence Stone. Baltimore: Johns Hopkins UP, 1976. 95–111.

Clark, William. "On the Bureaucratic Plots of the Research Library." *Books and the Sciences in History*. Ed. Marina Frasca-Spada and Nick Jardine. Cambridge: Cambridge UP, 2000. 190–206.

Clarke, Samuel. *The saints nosegay, or, A posie of 741 spirituall flowers both fragrant and fruitfull, pleasant and profitable*. London, 1642.

Clegg, Cynthia. *Press Censorship in Elizabethan England*. Cambridge: Cambridge UP, 1997.

Clouet, Francois. *Pierre Quthe*. 1562. Louvres, Paris. 8 May 2005. <http://www.wga.hu/index1.html>.

Cock, Edward. *The Pen's Triumph*. London, 1658.

Coldiron, A.E.B. "Public Sphere/Contact Zone: Habermas, Early Print, and Verse Translation." *Criticism* 46.2 (2004): 207–22.

Coles, William. *The art of simpling*. London, 1656.

Collins, Minta. *Medieval Herbals: The Illustrative Traditions*. London: British Library, 2000.

Comenius, Johann Amos. *The Gate of Tongues Vnlocked and Opened, or else, A Seminarie or Seed-Plot of all Tongues and Sciences*. London, 1633.

———. *Joh. Amos Commenii Orbis Sensualium Pictus*. London, 1659.

Comito, Terry. *The Idea of the Garden in the Renaissance*. Hassocks: Harvester, 1979.

———. "Renaissance Gardens and the Discovery of Paradise." *Journal of the History of Ideas* 32 (1971): 483–506.

Cook, Harold. "Physicians and Natural History." *Cultures of Natural History*. Ed. Nick Jardine, James Secord and Emma Spary. 1996. 91–105.

———. "Time's Bodies: Crafting the Preparation and Preservation of Naturalia." *Merchants and Marvels: Commerce, Science, and Art in Early Modern Europe*. Ed. Pamela H. Smith and Paula Findlen. New York: Routledge, 2002. 223–47.

Cooper, N. *Houses of the Gentry, 1480–1680*. New Haven and London: Yale UP, 1999.

Copeman, W.S.C. *Doctors and Disease in Tudor Times*. London: Dawson, 1960.

Copland, William. *A Boke of the Propreties of Herbes Called an Herball*. London, 1555. N. pag.

Coverdale, Miles. *Biblia the Byble, that is, the holy Scrypture of the Olde and New Testament, faithfully translated in to Englyshe*. London, 1535. N. pag.

Crane, Mary Thomas. *Framing Authority: Sayings, Self, and Society in Sixteenth-Century England*. Princeton: Princeton UP, 1993.

Creighton, Aileen Swafford. "William Turner as Reformer." Diss. U of Texas, 1960.

Cressy, David. "Books as Totems in Seventeenth-Century England and New England." *Libraries and Culture* 21.1 (1985): 92–106.

Cunningham, Andrew. "The Culture of Gardens." *Cultures of Natural History*. Ed. Nick Jardine, James Secord, and Emma Spary. Cambridge: Cambridge UP, 1996. 38–56.

Curtius, Ernst Robert. *European Literature and the Latin Middle Ages*. Trans. Willard R. Trask. London: Routledge & Kegan Paul, 1953.

Daston, Lorraine. "Type Specimens and Scientific Memory." *Critical Inquiry* 31.1 (2004): 153–82.

Daston, Lorraine and Katharine Park. *Wonders and the Order of Nature: 1150–1750*. New York: Zone, 1998.

De Bray, Lys. *The Art of Botanical Illustration*. Royston, Hertfordshire: Eagle, 2002.

De Bruyn, Frans. "The Classical Silva and the Generic Development of Scientific Writing in Seventeenth-Century England." *New Literary History* 32.2 (2001): 347–73.

De Jong, Erik. "Nature and Art: The Leiden Hortus as 'Musaeum'." *The Authentic Garden: A Symposium on Gardens*. Ed. Leslie Tjon Sie Fat and Erik de Jong, Erik. Leiden: Clusius Foundation, 1991. 37–61.

Digby, Kenelm. *The closet of the eminently learned Sir Kenelme Digbie Kt*. London, 1669.

Donne, John. "To Sir Henry Goodyere." *John Donne: Poetry and Prose*. Ed. Franke J. Warnke. New York: Random House, 1967. 161–63.

Durrfeld, Eike Barbara. "Terra Incognita: Toward a Historiography of Book Fastenings and Book Furniture." *Book History* 3 (2000): 305–13.

Eamon, William. *Science and the Secrets of Nature: Books of Secrets in Medieval and Early Modern Culture*. Princeton: Princeton UP, 1994.

Edwards, Ralph. *The Shorter Dictionary of English Furniture: From the Middle Ages to the Late Georgian Period*. London: Country Life, 1964.

Edwards, Richard. *The paradyse of daynty deuises*. London, 1576.

Eisenstein, Elizabeth. *The Printing Press as an Agent of Change: Communications and Cultural Transformations in Early-Modern Europe*. 2 vols. Cambridge: Cambridge UP, 1979.

Ek, Renske C. *Petrus Cadé Herbarium, 1566*. [Online exhibition.] Utrecht: Nationaal Herbarium Nederland, 2004. <http://www.bio.uu.nl/~herba/Cade/>.

Elsky, Martin. *Authorizing Words: Speech, Writing, and Print in the English Renaissance*. Ithaca: Cornell UP, 1989.

Elsner, John and Roger Cardinal, ed. *The Cultures of Collecting*. London, Reaktion, 1994.

England and Wales. *A proclamation deuised by the kinges hyghnes, with thaduise of his most honorable counsell, to auoide and abolish suche englishe bookes, as conteine pernicious and detestable errours and heresies*. London, 1546.

Ewan, Joseph. Foreword. *Hortus Botanicus: The Botanic Garden and the Book*. Compiled by Ian MacPhail. Morton Arboretum: 1972.

Eyler, Ellen C. *Early English Gardens and Garden Books*. Ithaca: Cornell UP for the Folger Shakespeare Library, 1963.

Fanger, Claire. "Things Done Wisely by a Wise Enchanter: Negotiating the Power of Words in the Thirteenth Century." April 25, 2005. <http://www.esoteric.msu.edu /Fanger.html>

Fat, Leslie Tjon Sie. "Clusius' Garden: A Reconstruction." *The Authentic Garden: A Symposium on Gardens*. Ed. Leslie Tjon Sie Fat and Erik De Jong. Leiden: Clusius Foundation, 1991. 3–12.

Fat, Leslie Tjon Sie and De Jong, Erik, ed. *The Authentic Garden: A Symposium on Gardens*. Leiden: Clusius Foundation, 1991.

Fenton, Roger. *A Perfume Against the Noysome Pestilence*. London, 1603.

Findlen, Paula. "Courting Nature." *Cultures of Natural History*. Ed. Nick Jardine, James Secord, and Emma Spary. Cambridge: Cambridge UP, 1996. 57–74.

———. "The Formation of a Scientific Community: Natural History in Sixteenth-Century Italy." *Natural Particulars: Nature and the Disciplines in Renaissance Europe*. Ed. Anthony Grafton and Nancy Siraisi. Cambridge, MA: Massachusettes Institute of Technology, 1999. 369–400.

———. *Possessing Nature: Museums, Collecting, and Scientific Culture in Early Modern Italy*. Berkeley: U of California P, 1994.

Finlay, Michael. *Western Writing Implements in the Age of the Quill Pen*. Wetheral, Carlisle, Cumbria: Plains, 1990.

Fish, Stanley. *Self-Consuming Artifacts: The Experience of Seventeenth-Century Literature*. Berkeley: U of California P, 1972.

Fissell, Mary. "Readers, Texts, and Contexts: Vernacular Medical Works in Early Modern England." *The Popularization of Medicine 1650–1850*. Ed. Roy Porter. The Wellcome Institute Series in the History of Medicine. London: Routledge, 1992. 72–96.

Fleming, Juliet. *Graffiti and the Writing Arts of Early Modern England*. Philadelphia: U of Pennsylvania P, 2001.

———. "Wounded Walls: Graffiti, Grammatology, and the Age of Shakespeare." *Criticism* 39.1(1997):1–30.

Florio, John. *Florios Seconde Frutes*. London, 1591.

Foucault, Michel. *The Order of Things: An Archaeology of the Human Sciences*. 1970. Trans. Alan Sheridan Smith. New York: Pantheon, 1971.

Francis, Jane. "The Kedermister Library: An Account of its Origins and a Reconstruction of its Contents and Arrangement." *Records of Buckinghamshire* 36 (1994): 62–85.

Frank, R.G. "The Physician as Virtuoso in Seventeenth-Century England." *English Scientific Virtuosi in the Sixteenth and Seventeenth Centuries*. Ed. B. Shapiro and R.G. Frank. Los Angeles: 1979. 57–114.

Frasca-Spada, Marina and Nick Jardine. "Introduction: Books and the Sciences." *Books and the Sciences in History*. Ed. Frasca-Spada, Marina and Nick Jardine. Cambridge: Cambridge UP, 2000. 1–10.

French, Roger. *Ancient Natural History*. London: Routledge, 1994.

Fulke, William. *A Goodly Gallerye with a Most Pleasaunt Prospect, into the Garden of Naturall Contemplation*. London, 1563.

Gaskell, Philip. *A New Introduction to Bibliography*. Oxford: Clarendon, 1974.

Genette, Gérard. *Paratexts: Thresholds of Interpretation*. Trans. Jane E. Lewin. Literature, Culture, Theory 20. Cambridge: Cambridge UP, 1997.

Gerard, John. *Catalogus Arborum*. London, 1597.

———. *Catalogus Arborum*. London, 1599.

———. *The Herball or Generall Historie of Plants*. 1597. The English Experience. 660A. Amsterdam: Theatrum Orbis Terrarum, 1974.

Getty Museum. "Bernard Palissy." April 24, 2005. <http://www.getty.edu/art/collections/bio/a867-1.html>

Getty Museum. "Joris Hoefnagel." April 22, 2005. <http://www.getty.edu/art/collections/bio/a992-1.html>

Giamatti, A. Bartlett. *The Earthly Paradise and the Renaissance Epic*. Princeton: Princeton UP, 1966.
Girouard, Mark. *Life in the English Country House: A Social and Architectural History*. New Haven: Yale UP, 1978.
Gloag, John. *English Furniture*. 5th ed. London: Black, 1965.
Goldgar, Anne. "Nature as Art: The Case of the Tulip." *Merchants and Marvels: Commerce, Science, and Art in Early Modern Europe*. Ed. Pamela H. Smith and Paula Findlen. New York: Routledge, 2002. 324–46.
Goody, Jack. *The Culture of Flowers*. Cambridge: Cambridge UP, 1993.
Great Britain. *Statutes of the Realm, 1225–1713*. Vol. 3. Microcard. 1822–1828.
Green, J. Reynolds. *A History of Botany in the United Kingdom From the Earliest Times to the End of the 19th Century*. London: Dent, 1914.
Greene, Edward Lee. *Landmarks of Botanical History*. Ed. Frank N. Egerton. 2 vols. Stanford: Stanford UP, 1983.
Guicciardini, Lodovico. *The Garden of Pleasure*. London, 1576.
Guilpin, Everard. *Skialetheia*. London, 1598.
Gunther, R.T. *Early British Botanists and their Gardens: Based on Unpublished Writings of Goodyer, Tradescant, and Others*. Oxford: Oxford UP, 1922.
Guppy, Henry. *The Evolution of the Art of Printing*. Manchester: Manchester UP, 1940.
Hadfield, Miles. *A History of British Gardening*. 1960. London: Penguin, 1986.
Halasz, Alexandra. *The Marketplace of Print: Pamphlets and the Public Sphere in Early Modern England*. Cambridge: Cambridge UP, 1997.
Harrison, Peter. *The Bible, Protestantism, and the Rise of Natural Science*. Cambridge: Cambridge UP, 1998.
Harrison, Robert Pogue. *Forests: The Shadow of Civilization*. Chicago: U of Chicago P, 1992.
Harrison, William. *The Description of England*. Ed. Georges Edelen. Folger Documents of Tudor and Stuart Civilization. Ithaca: Cornell UP, 1968.
Hartley, Dorothy. *Food in England*. 1954. London: Macdonald, 1979.
Hatton, R.G. *Handbook of Plant and Floral Ornament*. New York: Dover, 1960.
Havens, Earle. *Commonplace Books: A History of Manuscripts and Printed Books from Antiquity to the Twentieth Century*. New Haven, Conn.: Beinecke Rare Book and Manuscript Library, 2001.
Henrey, Blanche. *British Botanical and Horticultural Literature before 1800*. Vol. 1. London: Oxford UP, 1975.
Herbert, Edward. *The Life of Edward Lord Herbert of Cherbury Written by Himself With a Prefatory Memoir*. Edinburgh: Ballantyne, 1809.
Herron, Thomas. "'Goodly Woods': Irish Forests, Georgic Trees in Books 1 and 4 of Edmund Spenser's *Faerie Queene*." *Quidditas* 19 (1998): 97–122.
Hill, Ordelle G. *The Manor, the Plowman, and the Shepherd: Agrarian Themes and Imagery in Late Medieval and Early Renaissance English Literature*. Selinsgrove: Susquehanna UP, 1993.
Hill, Thomas. *The Gardeners Labyrinth*. 1577. London, 1594.

———. *A most briefe and pleasaunte treatise, teaching howe to dresse, sowe, and set a garden.* London, 1558.

Hobhouse, Penelope. *Plants in Garden History*. London: Pavilion, 1992.

Hodges, Richard. *The English Primrose: Far Surpassing al Others of this Kinde, that Ever Grew in Any English Garden*. London, 1644.

Hoeniger, David. "Konrad Gesner (1516–65) as a Student of Animals and Plants." *Familiar Colloquy: Essays Presented to Arthur Edward Barker*. Ed. Patricia Bruckmann. Ottawa: Oberon, 1978. 81–95.

Hoeniger, F. David. "How Plants and Animals Were Studied in the Mid-Sixteenth Century." *Science and the Arts in the Renaissance*. Ed. John W. Shirley and F. David Hoeniger. Washington, D.C.: Folger Shakespeare Library, 1985. 130–48.

———. *Medicine and Shakespeare in the English Renaissance*. Newark: U of Delaware P, 1992.

Hoeniger, F.D. and J.F.M. Hoeniger. *The Development of Natural History in Tudor England*. Charlottesville, VA: Folger Shakespeare Library, UP of Virginia, 1969.

———. *The Growth of Natural History in Stuart England from Gerard to the Royal Society*. Folger Shakespeare Library, UP of Virginia, 1969.

Holland, Henry. *Spiritual Preseruatiues Against the Pestilence*. London, 1593.

Hoskins, W.G. "The Rebuilding of Rural England 1570–1640." *Past and Present* 4 (1953): 44–59.

Howard, Maurice. *The Early Tudor Country House: Architecture and Politics, 1490–1550*. London: George Philip, 1987.

Howell, Thomas. *The arbor of amitie*. London, 1568.

Hunt, John Dixon. "'*Curiosities* to Adorn *Cabinets* and *Gardens*.'" *The Origins of Museums: The Cabinet of Curiosities in Sixteenth- and Seventeenth-Century Europe*. Ed. Oliver Impey and Arthur MacGregor. Oxford: Clarendon, 1985. 193–203.

Hunt, Tony. *Plant Names of Medieval England*. Woodbridge, Suffolk: Brewer, 1989.

Hunter, Andrew, ed. *Thornton and Tully's Scientific Books, Libraries, and Collectors: A Study of Bibliography and the Book Trade in Relation to the History of Science*. 4th rev. ed. Aldershot, Hampshire: Ashgate, 2000.

Impey, Oliver, and Arthur MacGregor, ed. *The Origins of Museums: The Cabinet of Curiosities in Sixteenth- and Seventeenth-Century Europe*. Oxford: Clarendon, 1985.

Jackson, Benjamin Daydon. "A Life of William Turner." *Libellus de Re Herbaria (1538) and The Names of Herbes (1548): Facsimiles*. London: Ray Society, 1965. 13–26.

James, N.D.G. *A History of English Forestry*. Oxford: Blackwell, 1981.

Jardine, Lisa. *Worldly Goods: A New History of the Renaissance*. New York: Norton, 1996.

Jardine, Nicholas, James Secord, and Emma Spary, ed. *Cultures of Natural History*. Cambridge: Cambridge UP, 1996.

Jayne, Sears. *Library Catalogues of the English Renaissance*. Godalming, Surrey: St. Paul's Bibliographies, 1983.

Jeffers, Robert H. *The Friends of John Gerard (1545–1612), Surgeon and Botanist.* Falls Village, CT: Herb Grower, 1967.

Johns, Adrian. *The Nature of the Book: Print and Knowledge in the Making.* Chicago: Chicago UP, 1998.

Johnson, Francis R. "Notes on English Retail Book-Prices, 1550–1640." *The Library*, fifth series, 5.2 (1950): 83–112.

Johnson, Thomas, ed. *The Herball or Generall Historie of Plantes*. By John Gerard. London, 1633. N. pag.

Johnston, Stanley H. *The Cleveland Herbal, Botanical, and Horticultural Collections: A Descriptive Bibliography of Pre-1830 Works from the Libraries of the Holden Arboretum, the Cleveland Medical Library Association, and the Garden Center of Greater Cleveland.* Kent, OH: Kent State UP, 1992.

———. *Cleveland's Treasures from the World of Botanical Literature.* Wilmington, Ohio: Orange Frazer, 1998.

Jones, Peter Murray. "Book Ownership and the Lay Culture of Medicine in Tudor Cambridge." *The Task of Healing: Medicine, Religion and Gender in England and the Netherlands, 1450–1800.* Ed. Hilary Marland and Margaret Pelling. 49–68.

Jones, Richard Foster. *The Triumph of the English Language: A Survey of Opinions Concerning the Vernacular from the Introduction of Printing to the Restoration.* Stanford: Stanford UP, 1953.

Jones, Whitney R.D. *William Turner: Tudor Naturalist, Physician and Divine.* London and New York: Routledge, 1988.

Kaden, Vera. *The Illustration of Plants and Gardens 1500–1850.* London: Victoria and Albert Museum, 1982.

Kaufmann, Thomas DaCosta. "Empiricism and Community in Early Modern Science and Art: Some Comments on Baths, Plants, and Courts." *Natural Particulars: Nature and the Disciplines in Renaissance Europe.* Ed. Anthony Grafton and Nancy Siraisi. Cambridge, MA: Massachusetts Institute of Technology, 1999. 401–17.

———. *The Mastery of Nature: Aspects of Art, Science, and Humanism in the Renaissance.* Princeton: Princeton UP, 1993.

Kaufmann, Thomas DaCosta and Virginia Roehrig Kaufmann. "The Sanctification of Nature. Observations on the Origins of *Trompe l'oeil* in Netherlandish Book Painting of the Fifteenth and Sixteenth Centuries." *The Mastery of Nature: Aspects of Art, Science, and Humanism in the Renaissance.* By Thomas DaCosta Kaufmann. Princeton: Princeton UP, 1993. 11–48.

Kemp, Theresa D. "The Lingua Materna and the Conflict over Vernacular Religious Discourse in Fifteenth-Century England." *Philological Quarterly* 78.3 (1999): 233–57.

Kenseth, Joy, ed. *The Age of the Marvelous.* Hanover, NH: Trustees of Dartmouth College, 1991.

Kibre, Pearl. "The Intellectual Interests Reflected in Libraries of the Fourteenth and Fifteenth Centuries." *Journal of the History of Ideas* 7.3 (1946): 257–97.

Kremers, Edward and George Urdang. *History of Pharmacy: A Guide and a Survey*. Philadelphia: Lippincott, 1940.

Kusukawa, Sachiko. "Illustrating Nature." *Books and the Sciences in History*. Ed. Marina Frasca-Spada and Nick Jardine. Cambridge: Cambridge UP, 2000. 90–113.

———. "Incunables and Sixteenth-Century Books." *Thornton and Tully's Scientific Books, Libraries, and Collectors: A Study of Bibliography and the Book Trade in Relation to the History of Science*. Ed. Andrew Hunter. 4th rev. ed. Aldershot, Hampshire: Ashgate, 2000. 115–63.

———. "Leonhart Fuchs on the Importance of Pictures." *Journal of the History of Ideas* 58.3 (1997): 403–27.

Langham, William. *The garden of health*. London, 1597.

Lawrence, George H.M. "Herbals: Their History and Significance." *History of Botany: Two Papers Presented at a Symposium held at the William Andrews Clark Memorial Library, December 7, 1963*. By George H.M. Lawrence and Kenneth F. Baker. Los Angeles: Clark Memorial Library, U of California, 1965. 1–35.

Lawson, William. *A nevv orchard and garden*. London, 1618.

Lechner, Joan Marie. *Renaissance Concepts of the Commonplaces*. New York: Pageant, 1962.

Lecoq, Anne-Marie. "The Garden of Wisdom of Bernard Palissy." *The Architecture of Western Gardens: A Design History from the Renaissance to the Present Day*. Ed. Monique Mosser and Georges Teyssot. Cambridge: Massachusetts Institute of Technology, 1991. 69–79.

Leslie, Michael and Timothy Raylor, ed. *Culture and Cultivation in Early Modern England: Writing and the Land.* Leicester: Leicester UP, 1992.

Lethbridge, J.B. "Anthological Reading and Writing in Tudor England." *Anthologies of British Poetry: Critical Perspectives from Literary and Cultural Studies*. Ed. Barbara Korte, Ralf Schneider and Stefanie Lethbridge. Amsterdam: Rodopi, 2000. 57–73.

Lipsius, Justus. *Tvvo bookes of constancie*. London, 1595.

Loach, Jennifer. *Edward VI*. Ed. George Bernard and Penry Williams. New Haven and London: Yale UP, 1999.

Loewenstein, Joseph. *The Author's Due: Printing and the Prehistory of Copyright*. Chicago: U of Chicago P, 2002.

Lyly, John. *Euphues. The Anatomy of Wyt*. London, 1578.

Lyte, Henry, trans. *A Niewe Herball*. By Rembert Dodoens. London, 1578.

McCutcheon, Elizabeth. *Sir Nicholas Bacon's Great House Sententiae*. English Literary Renaissance Supplements, No. 3. Claremont, CA: U of Hawaii and the Sir Francis Bacon Foundation and Library, 1977.

Mackenzie, Elizabeth. "The Growth of Plants: A Seventeenth-Century Metaphor." *English Renaissance Studies: Presented to Dame Helen Gardner in Honour of Her Seventieth Birthday.* Oxford: Clarendon, 1980. 194–211.

McKitterick, Rosamond. "Books and Sciences Before Print." Ed. Marina Frasca-Spada and Nick Jardine. *Books and the Sciences in History*. Cambridge: Cambridge UP, 2000. 13–34.

McLean, Antonia. *Humanism and the Rise of Science in Tudor England.* London: Heinemann, 1972.

MacQuoid, Percy and Ralph Edwards. *The Dictionary of English Furniture from the Middle Ages to the Late Georgian Period.* 2 vols. 2nd rev. ed. London: Antique Collector's Club, 1983.

McRae, Andrew. *God Speed the Plough: The Representation of Agrarian England, 1500–1660*. Cambridge: Cambridge UP, 1996.

Maihew, Edward. *A paradise of praiers.* London, 1613.

Man, John. *Gutenberg: How One Man Remade the World with Words*. New York: Wiley, 2002.

Manguel, Alberto. *A History of Reading*. London: HarperCollins, 1996.

Manwood, John. *Treatise of the Forests Lawes*. London, 1615.

Maplet, John. *A greene forest*. London, 1567.

Marchitello, Howard. "Garden Frisson." *Journal of Medieval and Early Modern Studies* 33.1 (2003): 143–77.

Markham, Gervase. *Certaine Excellent and New Inuented Knots and Mazes, For Plots for Gardens*. London, 1623.

———. *The English Housewife*. 1615. Ed. Michael R. Best. Kingston, Montreal: McGill-Queen's UP, 1986.

Marotti, Arthur F. "Malleable and Fixed Texts: Manuscript and Printed Miscellanies and the Transmission of Lyric Poetry in the English Renaissance." *New Ways of Looking at Old Texts*: *Papers of the Renaissance English Text Society*. Ed. W. Speed Hill. 159–73.

———. "The Transmission of Lyric Poetry and the Institutionalizing of Literature in the English Renaissance." *Contending Kingdoms: Historical, Psychological, and Feminist Approaches to the Literature of Sixteenth-Century England and France*. Ed. Marie-Rose Logan and Peter L. Rudnytsky. Detroit: Wayne State UP, 1991. 21–41.

Marvell, Andrew. "The Garden." *Selected Poetry and Prose*. Ed. Robert Wilcher. London: Methuen, 1986. 47–9.

Matthews, Leslie G. *History of Pharmacy in Britain.* Edinburgh: Livingstone, 1962.

Maunsell, Andrew. *The First Part of the Catalogue of English printed Bookes.* London, 1595.

Mercer, Eric. *Furniture, 700–1700*. London: Weidenfeld and Nicolson, 1969.

Merchant, Carolyn. *The Death of Nature: Women, Ecology, and the Scientific Revolution*. New York: Harper & Row, 1989.

Middleton, Thomas. *The puritaine*. London, 1607.

Milgate, W., ed. *The Satires, Epigrams and Verse Letters*. By John Donne. Oxford: Clarendon, 1967.

Moisan, Thomas. "Herrick, Hollar, and the Tradescants: Piecing Together a Seventeenth-Century Triptych." *Criticism* 43.3 (2001): 309–24.

Montaigne, Michel de. *Essays of Michael, seigneur de Montaigne in three books*. Trans. Charles Cotton. London, 1685–86.

Moran, James. *Printing Presses: History and Development from the Fifteenth Century to Modern Times*. Berkeley: U of California P, 1973.

Morton, A.G. *History of Botanical Science: An Account of the Development of Botany from Ancient Times to the Present Day*. London: Academic, 1981.

Moss, Ann. *Printed Commonplace-Books and the Structuring of Renaissance Thought*. Oxford: Clarendon, 1996.

Mullaney, Steven. "Strange Things, Gross Terms, Curious Customs: The Rehearsal of Cultures in the Late Renaissance." *Representations* 3 (1983): 40–67.

Newlands, Carole E. *Statius'* Silvae *and the Poetics of Empire*. Cambridge: Cambridge UP, 2002.

Newton, Richard C. "Making Books from Leaves: Poets Become Editors." *Print and Culture in the Renaissance: Essays on the Advent of Printing in Europe.* Ed. Gerald P. Tyson and Sylvia S. Wagonheim. Newark: U of Delaware P, 1986. 246–64.

Newton, Thomas, trans. *The Touchstone of Complexions*. By Levinus Lemnius. London, 1581.

Nutton, Vivian. "'A Diet for Barbarians': Introducing Renaissance Medicine to Tudor England." *Natural Particulars: Nature and the Disciplines in Renaissance Europe*. Ed. Anthony Grafton and Nancy Siraisi. Cambridge, MA: Massachusetts Institute of Technology, 1999. 275–94.

Ogilvie, Brian W. "Image and Text in Natural History, 1500–1700." *The Power of Images in Early Modern Science*. Ed. Wolfgang Lefèvre, Jürgen Renn, and Urs Schoepflin. Basel: Birkhäuser Verlag, 2003. 141–66.

———. "The Many Books of Nature: Renaissance Naturalists and Information Overload." *Journal of the History of Ideas* 64.1 (2003): 29–40.

———. *The Science of Describing: Natural History in Renaissance Europe*. Chicago: U of Chicago P, 2006.

Olmi, Giuseppe. "Science-Honour-Metaphor: Italian Cabinets of the Sixteenth and Seventeenth Centuries." *The Origins of Museums: The Cabinet of Curiosities in Sixteenth- and Seventeenth-Century Europe*. Ed. Oliver Impey and Arthur MacGregor. Oxford: Clarendon, 1985. 1–17.

Orlin, Lena Cowen. *Elizabethan Households*. Washington, D.C.: Folger Shakespeare Library, 1995.

———. "Gertrude's Closet." *Shakespeare Jahrbuch* 134 (1998): 44–67.

Otten, Charlotte F. *Environ'd with Eternity: God, Poems, and Plants in Sixteenth and Seventeenth Century England.* Lawrence, Kansas: Coronado, 1985.

Parkinson, John. *Paradisi in Sole Paradisus Terrestris*. 1629. The English Experience. 758. Amsterdam: Theatrum Orbis Terrarum, 1975.

———. *Theatrum botanicum*. London, 1640.

Pavord, Anna. *The Naming of Names: The Search for Order in the World of Plants*. London: Bloomsbury, 2005.

Peck, Linda Levy. "Building, Buying, and Collecting in London, 1600–1625." *Material London ca. 1600*. Ed. Lena Cowen Orlin. Philadelphia: U of Penn P, 2000. 268–89.

Pelling, Margaret. *The Common Lot: Sickness, Medical Occupations and the Urban Poor in Early Modern England.* London: Longman, 1998.

Pelling, Margaret and Charles Webster. "Medical Practitioners." Ed. Charles Webster. *Health, Medicine, and Mortality in the Sixteenth Century*. Cambridge: Cambridge UP, 1979. 165–235.

Pelling, Margaret (with Frances White). *Medical Conflicts in Early Modern London: Patronage, Physicians, and Irregular Practitioners, 1550–1640.* Oxford: Clarendon, 2003.

Perry, Jon Pearson. "Practical and Ceremonial Uses of Plants Materials as 'Literary Refinements' in the Libraries of Leonello d'Este and His Courtly Literary Circle." *La Bibliofilia* 91 (1989): 121–73.

Petroski, H. *The Book on the Bookshelf.* New York: Random House, 1999.

Pineas, Rainer. "William Turner's *Spirituall Physik.*" *Sixteenth Century Journal.* 14.4 (1983): 387–98.

Plant, Marjorie. *The English Book Trade: An Economic History of the Making and Sale of Books*. 3rd ed. 1939. London: Allen & Unwin, 1974.

Plat, Hugh. *Delightes for ladies to adorne their persons, tables, closets, and distillatories*. London, 1602. N. pag.

———. *Floraes paradise*. London, 1608.

———. *Floures of philosophie*. London, 1572.

Pomeroy, Elizabeth W. *The Elizabethan Miscellanies: Their Development and Conventions*. Berkeley: U of California P, 1973.

Pomian, Krzysztof. *Collectors and Curiosities: Paris and Venice, 1500–1800.* Cambridge: Polity, 1990.

Porter, Roy. "Introduction." *The Popularization of Medicine, 1650–1850*. Ed. Roy Porter. The Wellcome Institute Series in the History of Medicine. London: Routledge, 1992. 1–16.

Prest, John. *The Garden of Eden: The Botanic Garden and the Re-Creation of Paradise*. New Haven: Yale UP, 1981.

Puttenham, George. *The Arte of English Poesie*. 1589. Kent English Reprints. Kent, Ohio: Kent State UP, 1970.

Quest-Ritson, Charles. *The English Garden: A Social History*. London: Penguin, 2003.

Rainbowe, Edward. *A sermon preached at the funeral of the Right Honorable Anne, Countess of Pembroke, Dorset, and Montgomery*. London, 1677.

Raven, Charles E. *English Naturalists from Neckam to Ray: A Study of the Making of the Modern World.* Cambridge: Cambridge UP, 1947.

Reeds, Karen Meier. *Botany in Medieval and Renaissance Universities*. New York: Garland, 1991.

Rickman, Melissa L. "The Historical Context of Gerard's *Herball* and Johnson's Gerard: A Comparison." MLIS thesis U of Oklahoma Graduate College, 2001.

Rohde, Eleanour Sinclair. *The Old English Herbals*. London: Longmans, 1922.

Rota, Anthony. *Apart From the Text*. New Castle, DE: Oak Knoll, 1998.

Rovelstad, Mathilde V. "Two Seventeenth-Century Library Handbooks, Two Different Library Theories." *Libraries and Culture* 35 (2000): 540–56.

Sachs, Julius Von. *History of Botany (1530–1860)*. Trans. Henry E.F. Garnsey. Oxford: Clarendon, 1890.

Saenger, Michael. *The Commodification of Textual Engagements in the English Renaissance.* Aldershot, Hampshire: Ashgate, 2006.

Sanecki, Kay N. *History of the English Herb Garden*. London: Ward Lock, 1992.

Saunders, Gill. *Picturing Plants: An Analytical History of Botanical Illustration.* Berkeley, CA: U of California P, 1995.

Schofield, John. "The Topography and Buildings of London, ca. 1600." *Material London ca. 1600*. Ed. Lena Cowen Orlin. Philadelphia: U of Pennsylvania P, 2000. 296–321.

Shakespeare, William. *A midsommer nights dreame.* London, 1600.

———. *M. William Shak-speare: His True Chronicle Historie of the Life and Death of King Lear and his Three Daughters*. London, 1608.

Shapiro, Barbara. *A Culture of Fact: England, 1550–1720*. Ithaca: Cornell UP, 2000.

Sherman, William H. "On the Threshold: Architecture, Paratext, and Early Print Culture." Ed. Sabrina Alcorn Baron, Eric N. Lindquist, and Eleanor F. Shevlin. *Agent of Change: Print Culture Studies After Elizabeth Eisenstein*. Amherst, MA: U of Massachusetts P, 2007. 67–81.

Shevlin, Eleanor F. "'To Reconcile Book and Title, and Make 'em Kin to One Another': The Evolution of the Title's Contractual Functions." *Book History* 2.1 (1999): 42–77.

Singleton, Esther. *The Shakespeare Garden*. New York: Century, 1922.

Siraisi, Nancy G. *Medieval and Early Renaissance Medicine: An Introduction to Knowledge and Practice.* Chicago: U of Chicago P, 1990.

Skelton, John. *Garland of Laurells*. London, 1523.

Slack, Paul. "Mirrors of Health and Treasures of Poor Men: The Uses of the Vernacular Medical Literature of Tudor England." *Health, Medicine, and Mortality in the Sixteenth Century*. Ed. Charles Webster. Cambridge: Cambridge UP, 1979. 237–73.

Smith, Dinitia. "Has History Been Too Generous to Gutenberg?" *New York Times* 27 Jan. 2001. 1 April 2005. <http://www.nytimes.com/2001/01/27/arts/27PRIN.html>.

Smith, Pamela H. and Paula Findlen, ed. *Merchants and Marvels: Commerce, Science, and Art in Early Modern Europe*. New York: Routledge, 2002.

Spangenberg, Johann. *The Su[m] of Diuinitie*. Trans. Robert Hutten. London, 1548.

Spenser, Edmund. *The Faerie Queene*. Ed. A.C. Hamilton. Rev. ed. Harlow, England: Longman-Pearson, 2007.

Spufford, Margaret. *Small Books and Pleasant Histories: Popular Fiction and its Readership in Seventeenth-Century England.* London: Methuen, 1981.

Stearn, William Thomas. "Introduction: Agnes Arber's 'Herbals'." *Herbals: Their Origin and Evolution: A Chapter in the History of Botany, 1470–1670*. By Agnes Arber. 3rd ed. Cambridge: Cambridge UP, 1986. xxv–xxxii.

———. Preface. *Libellus de Re Herbaria (1538) and The Names of Herbes (1548): Facsimiles*. By William Turner. London: Ray Society, 1965. ix.

———. "The Use of Bibliography in Natural History." *Bibliography and Natural History*. Ed. Thomas R. Buckman. Lawrence, Kansas: U of Kansas Libraries, 1966. 1–26.

———. "William Turner's *Libellus*, 1538, and *Names of Herbes*, 1548." *Libellus de Re Herbaria (1538) and The Names of Herbes (1548): Facsimiles*. By William Turner. London: Ray Society, 1965. 3–9.

Steinberg, S.H. *Five Hundred Years of Printing*. Ed. John Trevitt. Rev. ed. London: The British Library and Oak Knoll, 1996.

Stewart, Alan. "The Early Modern Closet Discovered." *Representations* 50 (1995): 76–100.

Stow, John. *A Suruay of London*. London, 1598.

Streeter, B.H. *The Chained Library: A Survey of Four Centuries in the Evolution of the English Library*. London: Macmillan, 1931.

Strong, Roy. *The Artist and the Garden*. New Haven: Yale UP, 2000.

———. *The Renaissance Garden in England*. London: Thames and Hudson, 1979.

Summit, Jennifer. "Monuments and Ruins: Spenser and the Problem of the English Library." *ELH* 70 (2003): 1–34.

Swan, Claudia. "From Blowfish to Flower Still Life Paintings: Classification and its Images, circa 1600." *Merchants and Marvels: Commerce, Science, and Art in Early Modern Europe*. Ed. Pamela H. Smith and Paula Findlen. New York: Routledge, 2002. 109–36.

Swann, Marjorie. *Curiosities and Texts: The Culture of Collecting in Early Modern England*. Philadelphia: U of Pennsylvania P, 2001.

Thirsk, Joan. "Making a Fresh Start: Sixteenth-Century Agriculture and the Classical Inspiration." *Culture and Cultivation in Early Modern England: Writing and the Land*. Ed. Michael Leslie and Timothy Raylor. Leicester: Leicester UP, 1992. 15–34.

Thomas, Keith. *Man and the Natural World: A History of the Modern Sensibility*. New York: Pantheon, 1983.

Thornton, Dora. *The Scholar in his Study: Ownership and Experience in Renaissance Italy*. New Haven: Yale UP, 1997.

Thornton, Peter. *Seventeenth-Century Interior Decoration in England, France and Holland*. New Haven and London: Yale UP, 1979.

Tomasi, Lucia Tongiorgi. "Projects for Botanical and Other Gardens: A 16th-Century Manual." *Journal of Garden History* 3 (1983): 1–34.

Top, Alexander. *The Oliue Leafe: Or, Vniuersall Abce*. London, 1603.

Topsell, Edward. *History of Four-footed Beasts*. London, 1608.

Tory, Geoffroy. *Champ Fleury*. Trans. George B. Ives. 1927. New York: Klaus Reprint Corporation, 1967.

Tottell, Richard, ed. *Songes and sonettes*. London, 1557.

Townsend, George, ed. *The Acts and Monuments of John Foxe*. Vol. 5. 1846. New York: AMS, 1965.

Turner, William. *The huntyng and fyndyng out of the Romyshe foxe*. Antwerp, 1544. N. pag.

———. *The huntyng of the romyshe vuolfe*. Emden, 1555? N. pag.

———. *Libellus de Re Herbaria (1538) and The Names of Herbes (1548)*. London: Ray Society, 1965.

———. *A nevv booke of spirituall physik for dyuerse diseases of the nobilitie and gentlemen of Englande*. [Emden], 1555.

———. *A New Herball*. 1551. Ed. George T.L. Chapman and Marilyn N. Tweddle. Manchester: Mid Northumberland Arts Group and Carcanet, 1989.

———. *A New Herball by William Turner: Parts II and III*. 1562 and 1568. Ed. George T.L. Chapman, Frank McCombie and Anne Wesencraft. Cambridge: Cambridge UP, 1995.

Tusser, Thomas. *Fiue hundreth points of good husbandry vnited to as many of good huswiferie*. London, 1573.

———. *A hundreth good pointes of husbandrie*. London, 1557.

———. *A Hundreth good pointes of Husbandry*. London, 1570.

Visentini, Margherita Azzi. "A Model Humanist Garden: Villa Brenzone at Punta San Vigilio." *The Architecture of Western Gardens: A Design History from the Renaissance to the Present Day.* Ed. Monique Mosser and Georges Teyssot. Cambridge: Massachusetts Institute of Technology, 1991. 106–8.

Wabuda, Susan. *Preaching During the English Reformation*. Cambridge: Cambridge UP, 2002.

Wainwright, Clive. "The Library as Living Room." *Property of a Gentleman: The Formation, Organisation and Dispersal of the Private Library 1620–1920*. Ed. Robin Myers and Michael Harris. Winchester: St. Paul's Bibliographies, 1991. 15–24.

Wasserstein, A. "The Manuscript Tradition of Statius' *Silvae*." *Classical Quarterly* 47 (1953): 69–78.

Watt, Tessa. *Cheap Print and Popular Piety: 1550–1640*. Cambridge: Cambridge UP, 1991.

Wear, Andrew. "The Early Modern Debate about Foreign Drugs: Localism versus Universalism in Medicine." *The Lancet* 354 (1999): 149–51.

———. "Religious Beliefs and Medicine in Early Modern England." *The Task of Healing: Medicine, Religion and Gender in England and the Netherlands, 1450–1800*. Ed. Hilary Marland and Margaret Pelling. 145–69.

Webbe, William. *Discourse of English Poetrie*. London, 1586.

Weber, A.S. "Women's Early Modern Medical Almanacs in Historical Context ." *ELR* 33 (2003): 358–402.

Webster, Charles. *The Great Instauration: Science, Medicine and Reform 1626–1660*. London: Duckworth, 1975.

West, Susie. "Tales from the Book Room: Interdisciplinary Approaches to Country House Libraries and their Books." *Looking Forwards: The Country House in Contemporary Research and Conservation*. Ed. Leo Schmidt et al. Cottbus, Germany: Cottbus, 2001. 68–76.

Westerhoff, Jan. C. "A World of Signs: Baroque Pansemioticism, the Polyhistor and the Early Modern Wunderkammer." *Journal of the History of Ideas* 62.4 (2001): 633–50.

Wheathill, Anne. *A Handfull of Holesome (Though Homelie) Hearbs.* London, 1584.

Whitaker, Katie. "The Culture of Curiosity." *Cultures of Natural History*. Ed. Nick Jardine, James Secord, and Emma Spary. Cambridge: Cambridge UP, 1996. 75–90.

Whitford, Richard. *The Pomander of Prayer*. London, 1531.

Whitney, Isabella. *A Sweet Nosgay, or Pleasant Posye: Contayning a Hundred and Ten Phylosophicall Flowers.* London, 1573.

Williamson, T. *Polite Landscapes: Gardens and Society in Eighteenth-Century England.* Baltimore: Johns Hopkins UP, 1995.

Wright, Louis B. *Middle-Class Culture in Elizabethan England.* Chapel Hill, NC: U of North Carolina P, 1935.

Wroth, Mary. *The Countesse of Mountgomeries Urania*. London, 1621.

Zytaruk, Maria. " 'Occasional *Specimens*, Not Compleate *Systemes*': John Evelyn's Culture of Collecting." *Bodleian Library Record* 17 (2000): 185–212.

Index

Achinstein, Sharon 127
Adams, Thomas 42
Addison, Joseph 112
Alberti, Leon Battista 118
Albertus Magnus 9, 14n, 78
Aldrovandi, Ulisse 19, 27, 57
alphabetical order 70, 71, 92
Anatomy of Melancholy (Burton) 16
Anderson, Randall L. 5
anthologies; *see also* commonplaces; gardens of verse
  anthological characteristics of herbals ix, xii, 25, 46–9, 78–9
  compiler likened to bee 4, 49
  as composed of textual flowers 5, 79, 133
  Gerard's herbal as 80–82, 109
  recipe books as 126
  rhetoric of gathering in 47–8
apothecaries 15, 21, 46, 61, 125
Arber, Agnes 5n, 14, 22, 24, 31, 65, 78, 95
Arbor of amitie, The (Howell) 2, 3
Arbor of amorous deuises, The (Breton) 2
Aristophanes 100, 102
Arte of English Poesie, The (Puttenham) 95, 96, 120, 130
Ashworth, William B. 8, 13, 14
authorship; *see also* intellectual property
  authorial ownership 48–9, 76
  controversy over Gerard's herbal 75–83, 108–9
  early-modern collective 83
  emerging emphasis on original 18
  publishing for asserting authorial entitlement 84
  Virgilian literary model of 39–40

Bacon, Francis 18, 128, 134
Bacon, Nicholas 129
ballads, plagues linked to 127
Banckes's herbal 9, 22, 46
Barlow, H.M. 46
Bartholomaeus 78
Basse, William 126
Bauhin brothers 57
Beal, Peter 73n
Becon, Thomas 42, 127
Bel-vedére or, The garden of the muses (Bodenham) 4, 122
Bible
  Protestant emphasis on historicity of 58
  translation of 52–3, 56
bindings 125, 131
Bishop, John 1
Blair, Ann 40
Boccone, Paolo 32, 34
Bock, Jerome 25, 26, 57
Bocskay, Georg 32, 34, 35
Bodenham, John 4, 122
Boesky, Amy 4
book chests 9
Book of Common Prayer 53
Book of cookrye Very necessary for all such as delight therin, A (A.W.) 48
Book of Nature
  and authorial ownership 48
  botanical gardens in academic study of 16
  herbaria compared with 31, 38
  names seen as keys to unlocking 19
  Pepys's library as replica of 112
  Turner's conception of interpretation of 59
  word and thing mingled in 13
bookcases 114
books; *see also* anthologies; authorship; commonplaces; gardens of verse; herbals; herbaria; libraries; reading
  in advances in botany 14–15, 17
  as agents of disease 127
  as artificial galleries 122
  botanical and bibliographic interactions 1–12
  casual disposal of 125–6
  centrality in sixteenth-century botany 40

collecting 4
covers 9, 131, 133
as interiors 116–17
in kitchens 126
linguistic connections between plants and 10–11
in natural philosophy 40
pharmaceutical powers attributed to texts 125–6
places for 113–15, 131
plant materials in making and preservation of ix, 8–9
posies bridge gap between plants and 129–31
preserving 9, 125, 129
as rooms 131–2
seventeenth century severs connection between botanical discourse and 134
social standing in sixteenth-century England 111–13
in stillrooms 126
texts in Turner's work 39, 40–41, 44, 45
titles as means of marketing 112–13
troped as gardens 23, 113, 116–17, 122, 123–4
troped as other sorts of physical objects 113
botanical gardens
books identified as collections of plants 1
in botanical renaissance of the sixteenth century 6
catalogues of 28–9
herbaria as bookish counterparts to 29
at Padua 77
university 16, 28
Botanicum Monspeliense 19
botany; *see also* botanical gardens; medicinal herbalism; naming; plants
bibliographic and botanical interactions 1–12
collation of texts and nature in humanist 40
by correspondence 20–21, 31, 46–7
emergence of academic 7–8, 16
Gerard's contribution to 69–109
noble patronage for 16–17
seventeenth century severs connection between bookish discourse and 134
sixteenth-century renaissance of 6–7, 23, 39, 45, 69
social status of those involved in 44–5
texts in humanist 20–28
and theology 42
Turner's reformation in 39–67
Bredwell, Stephen 79, 81–2
Brenzone, Agostino 128
Breton, Nicholas 1–2
Brittons bowre of delights (Breton) 2
Brunfels, Otto
Herbarum Vivae Eicones 23, 24, 28
on naming 65
Reformation der Apotecken 57
in reformed church 57
Buch der Nature, Das (Megenberg) 78
Bulleins bulwarke 45, 83
Burton, Robert 16
Bushnell, Rebecca 5, 89n, 118, 130

cabinets 131, 132
cabinets of curiosities 33
Cadé, Petrus 30
Camerarius, Joachim 21, 27
candlelight 123
Cartwright, Thomas 47
Castel of Helth, The (Elyot) 53
catalogues 40, 70, 71–2
Catalogus arborum (Gerard) 29, 70, 71–3
Catullus 102–3
Cave, Terence 5
Cecil, William 17, 43, 44, 44n, 55, 69, 70
Cesalpino, Andrea 16n3, 57
Champ Fleury (Tory) xiii, xiv
Chartier, Roger 111, 119
Cibo, Gherardo 36
Cicero 93, 94, 97, 98
Cingulum 65–6
Clark, Peter 111, 114, 126
Clark, William 117
Clarke, Samuel 117
clasps 131
Clement, Claude 126
Clement, John 44
Clifford, Anne 130
Closet of the eminently learned Sir Kenelme Digbie Kt., The 127
closets 126–7, 131, 132

Clouet, François 29
Clusius, Carolus 16n3, 21, 24, 28, 57, 84
Coldiron, A.E.B. 111n
Coles, William 123, 127, 128
collecting; *see also* anthologies
  catalogues of collections 40, 72
  culture of 4–5
  Gerard's botanical aesthetic compared with that of 90–92
  herbaria as collectibles 34
  of medical prescriptions 21–2
  of medicinal plants 15–16
  plant names 14
  textual collection 4–5, 11–12, 14
College of Physicians 70, 77, 86
Colonna, Fabio 32
Comenius, Johann Amos 20, 123
Commentarii (Mattioli) 17, 21, 28
commonplaces
  compiler as bee analogy 49
  Gerard's herbal as 75–6, 93, 133
  herbals compared with 23, 28, 93
  medical prescription collections compared with 21–2
  and rhetoric of gathering 48
  rhetorical characteristics of 93
  "topics" in 73, 73n
commons, the 75
Co[m]parison betwene the olde learnynge & the newe, A (Turner) 52, 56
Cook, Harold 27
cookbooks 126
Copland, William 9
Copus, Guillelmus 16n2
Cordus, Euricius 57
Cordus, Valerius 57
correspondence, botanical 20–21, 31, 46–7
Coverdale, Miles 52, 62
covers 9, 131, 133
Crane, Mary Thomas 5, 48, 76, 82–3, 93
Cruydeboeck 76

De copia (Erasmus) 122
De Historia Stirpium (Fuchs) 24, 25, 27, 28
De Naturis Herbarum (Turner) 39
De Ratione (Erasmus) 129–30
De vegetabilibus (Albertus Magnus) 9
Debus, Allen 17, 22
Decembrio, Angelo 128–9
Delightes for ladies (Plat) 37
Denham, Henry 3
description, as trope 95–6
Description of England (Harrison) 7, 42, 45, 91
devotional books 1, 127
Digby, Kenelm 127
Dioscorides
  Gerard cites 100, 102, 103, 105
  Mattioli's commentary on 17, 18, 21
  Turner and 46, 54, 64
Discourse of English Poetrie (Webbe) 112
dissolution of the monasteries 54, 114, 125
Divine Herball, The (Adams) 22, 32, 42
Dodoens, Rembert
  Gerard acknowledges influence of 80
  Gerard and Priest's translation of 78–9
  Gerard cites xii
  Gerard corresponds with 81
  Gerard's herbal and work of 82
  images used by 24, 28
  as physician 16n3
Donne, John 2, 130

Eamon, William 48
Eden, Richard 19
Edwards, Richard 124
Edwards, Thomas 81
Eisenstein, Elizabeth 20, 76
Elyot, Thomas 53
empiricism, text-repudiating 134
English Husbandman, The (Markham) 120
Erasmus, Desiderius 8, 16n2, 122, 129–30
Evelyn, John 38

Faerie Queene (Spenser) 8–9, 10
Falconer, John 37, 44
Fenton, Roger 127
figurative language 94–7
filing cabinets 35
Findlen, Paula 5, 14–15, 18
Fisher, Cardinal 126
Fissell, Mary 119
Fiue hundreth points of good husbandry vnited to as many of good huswiferie (Tusser) 116
Fleming, Juliet 113, 131

fleurons 3, 9
Floraes paradise (Plat) 3, 116
florilegia 32
Florios seconde frutes (Florio) 124
Flour of Godly Praiers, The (Becon) 42
Floures of philosophie, The (Plat) 3, 117
flowers
  anthologies as textual 5, 79, 133
  christening 60
  in images in herbals 24
  ink made from 9
  poems compared with 32
  posies as pun on 131
  in private homes 7, 115–16
  texts compared with 31
food, preservation of 127
Foucault, Michel 13–14, 133
framing 82–3, 131
fruit
  books compared with 128–9
  gardeners experiment with varieties of 91
  grafting in farming of 10
  in images in herbals 24
  increasing number of varieties of 7
  posies associated with 130
  preserving 125
  in private gardens 115–16
Fuchs, Leonhart
  as court physician 16n3
  De Historia Stirpium 24, 25, 27, 28
  on physicians' knowledge of plants 16
  in reformed church 57
  Turner prefers apothecaries' names to those of 61
Fulke, William 120
Füllmaurer, Heinrich 25, 27

Galen 46, 54, 102
galleries
  posies on 129
  view of gardens from 120–23
Garcia da Orta 57
Garden of Health, The (Langham) 47–8, 79
Garden of pleasure, The (Guicciardini) 117
Garden of spiritual flowers, A (herbal) 22
Gardener's Labyrinth, The (Hill) 120, 121
gardens; *see also* botanical gardens; gardens of verse
  discontinuous experience of Tudor 118–19
  earliest English handbook on gardening 6
  in English botanical renaissance 6–7, 42, 115
  herbals construed as 23
  horticultural handbooks 4, 6
  as integrated with domestic interior 118
  monastery 42
  reading in 122
  as site for collection of plants 4
  studies associated with 127–9
  texts troped as 23, 113, 116–17, 122, 123–4
  viewed from galleries 120–23
  visual and olfactory relationship to house 119–24
  as walled and private 72, 117, 118, 131
gardens of verse
  and book as room metaphor 131
  and books as gardens trope 23
  botanical collection compared with 5, 49
  discontinuous experience of 118–19
  Gerard's herbal and 89, 90, 108
  herbaria compared with 32, 38
  as model for collection of texts 11
  and preservation of plants and books 125
  social status lacking for 111
Gascoigne, George 1
Gate of tongues vnlocked and opened, The (Comenius) 20
gathering, rhetoric of 47–8, 82–3
Gerard, John 69–109
  academic post sought by 70
  anthological literature's influence on 11
  Cecil as employer of 17, 69
  as curator of College of Physicians' physic garden 70, 77
  decline of reputation of 70
  in English botanical renaissance 6
  herbalists of his acquaintance 81
  lack of academic credentials of 70, 71–2
  and Priest's translation of Dodoens 78, 79–80
  reputation of 70, 78
  as surgeon 69, 70
  Turner compared and contrasted with 69–70

works (*see also Herball or Generall Historie of Plantes, The*)
Catalogus arborum 29, 70, 71–3
Gesner, Conrad 21, 27, 45, 46, 57
Gesta Grayorum (Bacon) 128
Ghini, Luca 15, 21, 36, 37
Gibson, Thomas 44
Gloag, John 9
Golden Rod 90–91
Goodly Gallerye with a Most Pleasaunt Prospect, A (Fulke) 120
Goody, Jack 2
Goodyere, Henry 130
grafting 10
Grafton, Anthony 119
Great Rebuilding 113
Greene forest, A (Maplet) 3, 3
Grete Herball, The 44, 49, 78, 116
Guicciardini, Lodovico 117
Guilpin, Everard 125

Handfull of holesome (though homelie) hearbs, A (Wheathill) 88, 119
Harrison, Peter 14n, 17, 20, 41, 52, 57, 62
Harrison, William 7, 42, 45, 91
Havens, Earle 126
Helpe to Memory and Discourse, A (Basse) 126
Herbal for the Bible (Newton) 42
Herball or Generall Historie of Plantes, The (Gerard)
as ambivalent about place of fiction in herbalism 103–8
anthological character of 80–82, 109
author's presence in 73–5, 83
authorship of 75–83, 108–9
chapter on fictional plants in 105–7
as close of botanical renaissance 69
commonplacing method in 75–6, 93, 133
contemporary view of 70n2
description of plants in 88–99
division into books and chapters 73
entrenched in history of poetry xii
on flowers 116
on gardens as ornaments and necessities 117
as "gathered" by Gerard 83
Gerard does not claim sole authorship of 80
on the hyacinth x–xi, xii–xiv, 9
on material connection of plants and books ix
on naming 62–3, 83–8, 108–9
oral verse tradition embedded in 92–3
ordering principle of 92
plagiarism attributed to 11, 46, 48, 78, 80, 82, 108
on plants as form of capital 76–7
poetry compared with descriptive techniques of 92, 94, 96, 97, 98, 109
poets cited by 99–104, 109
prefatory poems in 74, 75
as representation of Gerard's garden 72–3
second edition "To the Reader" on nomenclature 18
on sharing plants 21
similitude in 94–7
sources for 81–2
as textual commons 75
threefold task of 80
translating the visual into the verbal 97–8
two-page spreads from *x–xi,* 101
vivacity of description in 93–4
on wood for woodcuts 9
herbals; *see also Herball or Generall Historie of Plantes, The* (Gerard); *New Herball, A* (Turner); *and others by name*
anthological characteristics of ix, xii, 25, 46–9, 78–9
collaboration in production of 25
and commonplace tradition 23, 28, 93
disjunctive reading experience of 119
dried leaves added to 37
end of age of 69
as gardens 23
images in 23–5
increase in production of 40
in kitchens 126
nomenclature for 22–3
ordering principles of 92
poetic material in 25, 134
printed 22–8

rhetorical techniques of poets compared with those of 92
similitude in 94–5
usual scholarly approach to 5–6
variety as characteristic of 92
Herbals (Arber) 5n
herbaria 28–38
animal counterparts of 33
book form replaced by filing cabinet 34–5, 134
of Cadé 30
as collectibles 34
contemporary 35
early 36–7
first published evidence of 37
Hoefnagel's paintings compared with 32–3
as hybrid of plants and books 29, 31, 38, 133
invention of 36
as narrowing gap between representation and represented 32, 38
of Platter 33–4
practical uses of 31–2
production process for 35–6
of Quthe 29–31
restoring appearance of life of plants in 32
revolutionary character of 31
shelving with others 31
simpler methods for preserving plants 37
of Turner 39
various names for 29, 31
Herbarum Vivae Eicones (Brunfels) 23, 24, 28
Herbert, Edward, Lord 37–8
herbwives 15, 46, 86
Hesiod 102
Hill, Thomas 6, 120, 121
Historia general y natural (Oviedo) 19
Hoefnagel, Joris 32–3, 34, 35
Homer 102–4
Horace 100, 102
horticultural handbooks 4, 6
hortus hyemalis; see also herbaria
as term for herbarium 29, 31
hortus siccus; see also herbaria
as term for herbarium 29, 133
Hospitall for the Diseased, An (Cartwright) 47
How to dresse, sowe, and set a garden (Hill) 6
Howell, Thomas 2, 3
humanism
botanical metaphors used by 5
class position sought by humanists 76
collation of texts and nature in botany of 40
collection and redeployment of textual fragments in 48
gathering and framing in 82–3
ideal of living text 18, 23
naming in humanist herbalism 17–20
noble patronage for natural historians 16
revivification of classical texts by 8
texts in botany of 20–28
Hundreth good pointes of husbandrie (Tusser) 7, 8, 130
Huntyng & fyndyng out of the Romishe fox, The (Turner) 43, 52–3
hyacinth *x–xi,* xii–xiv
Hyacinthus xii, xiv
Hystoire of the Western Indies, The (Oviedo) 19

Impey, Oliver 4–5
ink 9, 133
intellectual property
plagiarism 11, 46, 48–9, 78, 80, 82, 108
varying reifications of 76

Jardine, Lisa 4, 119
Jayne, Sears 111
Johns, Adrian 24
Johnson, Thomas 7, 78
Jones, Richard 1
Jonson, Ben 71

Kandel, David 25, 26
King Lear (Shakespeare) xii, 6
kitchens 126
Kreuterbuch (Bock) 25, 26
Kusukawa, Sachiko 17

ladyslipper 60
Langham, William 48, 79
Lawson, William 120–21

Leland, John 124n
Lemnius, Levinus 115–16
Leoniceno, Niccolò 17
Libellus de re herbaria novus (Turner)
  on false names 63
  Gerard's *Catalogus arborum* compared with 70
  Names of Herbes as expanded translation of 54
  on Narcissus 59
  originality of 46
  publication of 41
  on study of plants as widespread 45
  title's significance 50
libraries
  domestic 112, 114–15
  integration with gardens 128
  pharmacies likened to 126
  plant materials in 129
  stillrooms compared with 126
linguistic nationalism 51
Linnaeus, Carolus 18
Lipsius, Justus 122n, 123
literacy 15
L'Obel, Matthias 16n3, 24, 28, 78, 81, 82, 83
Loewenstein, Joseph 76
Lorkyn, Thomas 114
Luis of Granada 75–6
Lusitanus, Amatus 37
Luther, Martin 57–8
Lyte, Henry 76, 80

MacGregor, Arthur 4–5
MacQuoid, Percy 113
magazines 131
Maihew, Edward 124
Maplet, John 3, 3
Marchitello, Howard 38
Markham, Gervase 120
Marotti, Arthur F. 21n, 126
Martial 103
Marvell, Andrew 124
Mattioli, Pier Andrea 16n3, 17–18, 21, 28, 32, 91, 105
measurement, similitude preferred to 95
medicinal herbalism; *see also* botany; herbals; physicians
  anthological textual productions of 25
  apothecaries 15, 21, 46, 61, 125
  collecting prescriptions 21–2
  history of Renaissance 15–17
  lay practitioners permitted 54–5
  literacy as credential in 15
  pharmaceutical powers attributed to texts 125–6
  physic gardens 21, 28, 91, 117
  stillrooms for manufacturing and storing remedies 125, 127
  Turner and 15, 39
Megenberg, Konrad von 78
Melanchthon, Philip 57
memory
  herbalists' reliance on 92–3
  posies as aids to 129–30
Meyer, Albrecht 25, 27
Middleton, Thomas 121
Midsommer nights dreame (Shakespeare) 49
Moisan, Thomas 25n
monastic dissolution 54, 114, 125
Montaigne, Michel de 33–4, 121, 122, 124, 128
More, Sir William 127
Morgan, Hugh 21, 46
Moss, Ann 5, 49

Names of Herbes (Turner)
  as catalogue 40
  contents of 50
  and Duke of Somerset's patronage 41
  as early book on British plants 46
  on experience in naming 63
  on false names 62
  foreign language names in 51n3
  as "gathering" 47
  on lack of names for plants 58
  Latin herbal referred to by 39
  on misidentification of plants 63–4
  multilingual program of 50
  multiple-name coining in 65–6
  on plants in Syon park 43
  as response to lay herbalism 55
  title's significance 50
  Turner's scriptural translation and 56
  as vernacular translation 50, 54
naming
  Adamic nomenclature 58, 62

in early modern botany xii, 11
Gerard's *Herball* on 62–3, 83–8, 108–9
of herbals 22–3
in humanist herbalism 17–20
imaginative aspect of 66
increasing rigor in 134
linguistic variation within England and 61–2
theological factors in 60, 62
in Turner's work 49–51, 58–67, 98
Narcissus 59, 60
natural history; *see also* botany
approaches to study of history of 13–14
Christian reformation and 57–8
literary mindset for understanding early modern 27
natural-historical collections 5
noble patronage for 16–17
nomenclature problems in 18
natural philosophy, texts in 40
New Booke of Spirituall Physick (Turner) 53, 55
New Herball, A (Turner)
English plants as focus of 40
"gathering" in production of 48
on hyacinth juice as adhesive ix, 9
naming in 50
originality claimed for 46
plagiarism accusations against 49
publication of three volumes of 43
sources listed for 83
on translation 50–51
variety of contributors to 47
as vernacular 45, 77
wide audience for 45
New Orchard and Garden (Lawson) 121
Newton, Thomas 42
Nicolls, Thomas 24n
Norton, John 78, 82
Novum Organon (Bacon) 18

Ogilvie, Brian 5n, 6n3, 19
On the Errors of Pliny and Others in Medicine (Leoniceno) 17
Ortus Sanitatis 78
Ovid xii–xiii, 102
Oviedo, Gonzalo Fernández de 19–20
Owen, George 44

Padua 77
Palissy, Bernard 33, *36,* 128
paper
as disease carrier 127
in herbarium production 36
rag pulp for 9
Paradise of praiers and meditations, A (Maihew) 124
Paradisi in Sole (Parkinson) 4, 69n, 117
Paradyse of daynty deuises, The (Edwards) 124
Parkinson, John 4, 7, 69n, 117, 120
paste, hyacinth juice for making ix
Pavord, Anna 6n3
Pelling, Margaret 21–2, 25
pens 9
Pepys, Samuel 112
Perfume Against the Noisome Pestilence (Fenton) 127
peristephanon 2
Perry, Jon Pearson 129
Pharmacopoeia Londinensis 77
physic gardens 21, 28, 91, 117
physicians
College of Physicians 70, 77, 86
combining posts of priest and 42
lay herbalists permitted 54–5
and plant collectors 15–16
and poets set on par in Basse's *A Helpe to Memory and Discourse* 126
Turner on ignorance of plants of 58–9
vernacular medical books opposed by 53–4, 77
piracy, literary 48–9
plagiarism 11, 46, 48–9, 78, 80, 82, 108
plagues, ballads linked to 127
Plant, Marjorie 111
Plantin, Christophe 28, 78
plants; *see also* botany; flowers; fruit; gardens
in book preservation 9
botanical and bibliographic interactions 1–12
collecting 4, 11–12
domestic uses of 124–5
expansion of number of specimens available 40
in libraries 129

linguistic connections between books and 10–11
materials in making of books ix, 8–9
names for textual collections drawn from 2
places for 115–19
posies bridge gap between books and 129–31
preserving 125
saints' names for 60
sending material in correspondence 20–21

Plat, Hugh 3, 37, 116, 117
Platter, Felix 33–4
Plautus 102
Pliny 17, 40, 46, 54, 99, 102, 108
plot 11
poetry; *see also* gardens of verse
botanical metaphors in poetic anthologies xii
as diversion during winter 127
Gerard's descriptive techniques compared with 92, 94, 96, 97, 98, 109
Gerard's herbal cites poets 99–104, 109
Gerard's herbal compares pleasures of plants with 89
herbals' composition resembles that of 25, 134
imaginative aspect of botanical naming and 66
physicians and poets set on par in Basse's *A Helpe to Memory and Discourse* 126
posies 129
restoring from page by reading aloud 32
Shakespeare on 49
Tottel's *Songes and sonnettes* 8
verse in botanical handbooks 4
women's botanical verse 88

Pomeroy, Elizabeth 119n3
porches 132
Porter, Roy 126
posies 129–31
Preseruative, or Triacle, Agaynst the Poyson of Pelagius, A (Turner) 55
preservation
of books 9, 125, 129
of food 127
of plant products 125

Prest, John 29
Priest, Robert 78, 79–80, 81, 82
printing
arrival as cultural industry in England 8
in botany's development 14
casual disposal of printed materials 125–6
incorporating print by allying it with other forms of communication 113
increase in number of books due to 113, 114
links between plants and books in 9–10, 133
low economic status of printed works 112
in medical herbalism 17
printed herbals 22–8
printed sheets as wall hangings 130
title pages 131

Prudentius 2, 102
Puritan, The (Middleton) 121
Puttenham, George 95, 96, 120, 130

Quthe, Pierre 29–31, 33

rarity, aesthetic of 90–91
Raven, Charles E. 6n3, 50, 79
reading
in bed 124
as discontinuous and disjunctive 119
in gardens 122
likened to meandering through garden 121–2
literacy 15
natural light for indoor 122–4
Palissy associates processing of plants with 128

recipe books 126
reed pens 9
Reeds, Karen Meier 20, 21, 23, 58
Reformation der Apotecken (Brunfels) 57
Rhetorica Ad Herennium (Cicero) 93, 94, 97, 98
Rhetorica ecclesiastica (Luis of Granada) 75–6
Rich, John 21, 46
Rickman, Melissa L. 70n2
Robin, John 81, 84
Ruel, Jean 61

Sachs, Julius von 6

St. John's Wort 60
saints' names 60
Saints nosegay, The (Clarke) 117
Schofield, John 112, 120
Sententiarum Flores (Turner) 39
Seymour, Edward, Duke of Somerset 16, 41, 43, 44
Shakespeare, William xii, 6, 49
Sherman, William H. 131–2
Shrewsbury, Elizabeth Hardwick, Countess of 124
similitude 94–7
Skialetheia (Guilpin) 125
Smale handfull of fragrant flowers (Breton) 1–2
Solomon, King 45
Somerset, Edward Seymour, Duke of 16, 41, 43, 44
Songes and sonnettes (Tottel) 8
Speckle, Veit Rudolf 25, *27*
Spenser, Herbert 8–9, 10
Spieghel, Adrian van 35–6
Spiritual Nosegay, The (Turner) 41, 55
Stationer's Company 8
Stearn, William T. 14, 46
Stewart, Alan 126–7, 132
stillrooms 125–7, 129
Stow, John 45, 118n
studies
  books likened to 132
  gardens associated with 127–9
  stillrooms compared with 126, 129
Swann, Marjorie 5
Sweet Nosegay (Whitney) 10, 88
Sweet posie for Gods saints, A (herbal) 22
Sweet William 60
Syon House 42–3

Tabernaemontanus 24, 78, 81
taxonomy 18
texts *see* books
Theatrum botanicum (Parkinson) 7
Theocritus 100
Theophrastus 17, 103, 108
Thorney, Thomas 74, 75
Thornton, Dora 128–9
Thornton, Peter 114, 120, 129
title pages 131
Tory, Geoffrey xiii, xiv
Tottel, Richard 8, 119n3
Tradescant's Ark 72
translation
  of Bible 52–3, 56
  in Turner's work 50–54
tulips 4, 11
Turner, Peter 47
Turner, William 39–67
  academic post sought by 70
  audience of 45
  Bible translation edited by 56
  biography of 41–5
  as church reformer 39, 41, 43, 45, 52, 55–7, 70
  correspondence of 46–7
  Duke of Somerset as patron of 16, 41, 44
  in English botanical renaissance 6
  English names as focus of 51
  as "father of English botany" 45, 70, 70n3, 133
  field trips of 8, 39, 41
  first publication of plants as decisive development 39
  Gerard acknowledges influence of 80
  Gerard compared and contrasted with 69–70
  Gerard draws on ix
  Ghini as teacher of 15, 37
  herbarium of 37, 39
  increasing size of botanical books of 7, 39–40
  naming in work of 49–51, 58–67, 98
  originality of 46, 62
  as physician and priest 42, 54
  on physicians' ignorance of plants 58–9
  Reformation's influence on 11
  reputation of 45, 70
  Syon House gardens available to 42–3
  texts in work of 39, 40–41, 44, 45
  theological work's influence on botanical work 55–7
  translation in work of 50–54
  women consulted by 45
  works (*see also Libellus de re herbaria novus; Names of Herbes*)
    A Co[m]parison betwene the olde learnynge & the newe 52, 56

De Naturis Herbarum 39
The Huntyng & fyndyng out of the Romishe fox 43, 52–3
New Booke of Spirituall Physick 53, 55
A Preseruative, or Triacle, Agaynst the Poyson of Pelagius 55
Sententiarum Flores 39
The Spiritual Nosegay 41, 55
Tusser, Thomas 7, 116, 130
Tyndale, William 52

variety 91–2
vernacular literature 11, 25, 45, 52–4, 77, 111–12
verse *see* poetry
vestiarian controversy 43
Villa Brenzone 128
Virgil xii, 39–40, 100, 102–3
vulgar tongue 78

Watt, Tessa 8n, 56n6, 113, 115, 130, 133
Webbe, William 112
Weiditz, Hans 23, 24
Wendy, Thomas 44
Wentworth, Lord Thomas 41
West, Susie 114, 115
Wheathill, Anne 88, 118
Whitney, Isabella 10, 88
windows 120, 122
women
botanical verse of 88
herbwives 15, 46, 86
lay herbalists permitted 54–5
naming plants after and by 85–8
in traditional medicine 15
Turner consults 45
woodcuts 9, 23, 24, 28, 32, 48, 78
wormwood 9, 63
Wotton, Edward 44
Wroth, Mary 2

Zytaruk, Maria 38

For Product Safety Concerns and Information please contact our EU
representative GPSR@taylorandfrancis.com
Taylor & Francis Verlag GmbH, Kaufingerstraße 24, 80331 München, Germany

www.ingramcontent.com/pod-product-compliance
Ingram Content Group UK Ltd.
Pitfield, Milton Keynes, MK11 3LW, UK
UKHW020205200926
475702UK00004B/622
* 9 7 8 0 7 5 4 6 6 5 8 6 1 *